D0439692

Linear Algebra

by
Steven A. Leduc

Series Editor
Jerry Bobrow, Ph.D.

Wiley Publishing, Inc.

Acknowledgement:
My continued thanks to Doug Lincoln, Michele Spence, and Jerry Bobrow for their faith and friendship and for the opportunity to once again contribute to this series. My appreciation also goes to Jenn Nemec for her meticulous typesetting.

Dedication:
This work is dedicated to my mother, Deborah my sister, Karen and my niece, Chelsea

Publisher's Acknowledgments
Cover photo by Stephen Johnson/Tony Stone Images

Production
Wiley Publishing, Inc., Indianapolis Composition Services

CliffsNotes™ *Linear Algebra*

Published by:
Wiley Publishing, Inc.
909 Third Avenue
New York, NY 10022
www.wiley.com

It is assumed that at this point in your mathematical education, you are familiar with the basic arithmetic operations and algebraic properties of the **real numbers**, the set of which is denoted **R**. Since the set of reals has a familiar geometric depiction, the number line, **R** is also referred to as the **real line** and alternatively denoted $\mathbf{R}^1$ ("R one").

The Space $\mathbf{R}^2$

Algebraically, the familiar x-y plane is simply the collection of all pairs (x, y) of real numbers. Each such pair specifies a **point** in the plane as follows. First, construct two copies of the real line—one horizontal and one vertical—which intersect perpendicularly at their origins; these are called the **axes**. Then, given a pair (x_1, x_2), the **first coordinate**, x_1, specifies the point's horizontal displacement from the vertical axis, while the **second coordinate**, x_2, gives the vertical displacement from the horizontal axis. See Figure 1. Clearly, then, the order in which the coordinates are written is important since the point (x_1, x_2) will not coincide—generally—with the point (x_2, x_1). To emphasize this fact, the plane is said to be the collection of *ordered* pairs of real numbers. Since it takes two real numbers to specify a point in the plane, the collection of ordered pairs (or the plane) is called **2-space**, denoted $\mathbf{R}^2$ ("R two").

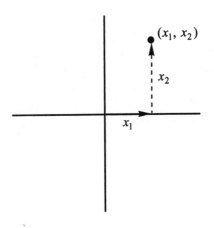

■ Figure 1 ■

$\mathbf{R}^2$ is given an algebraic structure by defining two operations on its points. These operations are **addition** and **scalar multiplication**. The sum of two points $\mathbf{x} = (x_1, x_2)$ and $\mathbf{x}' = (x_1', x_2')$ is defined (quite naturally) by the equation

$$\mathbf{x} + \mathbf{x}' = (x_1, x_2) + (x_1', x_2') = (x_1 + x_1', x_2 + x_2')$$

and a point $\mathbf{x} = (x_1, x_2)$ is multiplied by a **scalar** c (that is, by a real number) by the rule

$$c\mathbf{x} = c(x_1, x_2) = (cx_1, cx_2)$$

Example 1: Let $\mathbf{x} = (1, 3)$ and $\mathbf{y} = (-2, 5)$. Determine the points $\mathbf{x} + \mathbf{y}$, $3\mathbf{x}$, and $2\mathbf{x} - \mathbf{y}$.

The point $\mathbf{x} + \mathbf{y}$ is $(1, 3) + (-2, 5) = (-1, 8)$, and the point $3\mathbf{x}$ equals $3(1, 3) = (3, 9)$. Since $-\mathbf{y} = (-1)\mathbf{y} = (2, -5)$,

$$2\mathbf{x} - \mathbf{y} = 2\mathbf{x} + (-\mathbf{y}) = 2(1, 3) + (2, -5)$$
$$= (2, 6) + (2, -5) = (4, 1)$$

By defining $\mathbf{x} - \mathbf{x}'$ to be $\mathbf{x} + (-\mathbf{x}')$, the difference of two points can be given directly by the equation

$$\mathbf{x} - \mathbf{x}' = (x_1, \ x_2) - (x_1', \ x_2') = (x_1 - x_1', \ x_2 - x_2')$$

Thus, the point $2\mathbf{x} - \mathbf{y}$ could also have been calculated as follows:

$$2\mathbf{x} - \mathbf{y} = 2(1, \ 3) - (-2, \ 5) = (2, \ 6) - (-2, \ 5)$$
$$= (2 - (-2), \ 6 - 5) = (4, \ 1) \quad \blacksquare$$

Vectors in $\mathbf{R}^2$. A **geometric vector** is a directed line segment from an **initial point** (the **tail**) to a **terminal** or **endpoint** (the **tip**). It is pictured as an arrow as in Figure 2.

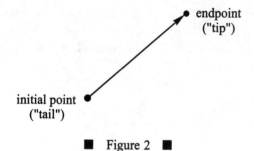

endpoint ("tip")

initial point ("tail")

■ Figure 2 ■

The vector from point $\mathbf{a}$ to point $\mathbf{b}$ is denoted $\mathbf{ab}$. If $\mathbf{a} = (a_1, a_2)$ is the initial point and $\mathbf{b} = (b_1, b_2)$ is the terminal point, then the signed numbers $b_1 - a_1$ and $b_2 - a_2$ are called the **components** of the vector $\mathbf{ab}$. The first component, $b_1 - a_1$, indicates the horizontal displacement from $\mathbf{a}$ to $\mathbf{b}$, and the second component, $b_2 - a_2$, indicates the vertical displacement. See Figure 3. The components are enclosed in parentheses to specify the vector; thus, $\mathbf{ab} = (b_1 - a_1, b_2 - a_2)$.

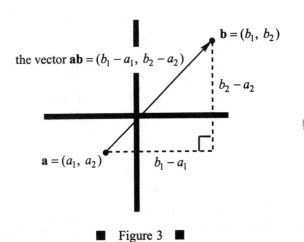

the vector $\mathbf{ab} = (b_1 - a_1, \ b_2 - a_2)$

$\mathbf{b} = (b_1, \ b_2)$

$b_2 - a_2$

$\mathbf{a} = (a_1, \ a_2)$

$b_1 - a_1$

■ Figure 3 ■

Example 2: If $\mathbf{a} = (4, 2)$ and $\mathbf{b} = (-5, 6)$, then the vector from $\mathbf{a}$ to $\mathbf{b}$ has a horizontal component of $-5 - 4 = -9$ and a vertical component of $6 - 2 = 4$. Therefore, $\mathbf{ab} = (-9, 4)$, which is sketched in Figure 4.

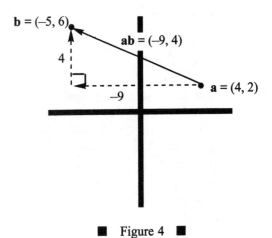

$\mathbf{b} = (-5, 6)$

$\mathbf{ab} = (-9, 4)$

4

-9

$\mathbf{a} = (4, 2)$

■ Figure 4 ■

Example 3: Find the terminal point of the vector $\mathbf{xy} = (8, -7)$ if its initial point is $\mathbf{x} = (-3, 5)$.

Since the first component of the vector is 8, adding 8 to the first coordinate of its initial point will give the first coordinate of its terminal point. Thus, $y_1 = x_1 + 8 = -3 + 8 = 5$. Similarly, since the second component of the vector is -7, adding -7 to the second coordinate of its initial point will give the second coordinate of its endpoint. This gives $y_2 = x_2 + (-7) = 5 + (-7) = -2$. The terminal point of the vector $\mathbf{xy}$ is therefore $\mathbf{y} = (y_1, y_2) = (5, -2)$; see Figure 5.

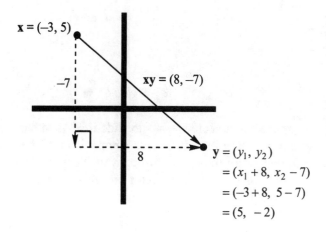

■ Figure 5 ■

Two vectors in $\mathbf{R}^2$ are said to be **equivalent** (or **equal**) if they have the same first component and the same second component. For instance, consider the points $\mathbf{a} = (-1, 1)$, $\mathbf{b} = (1, 4)$, $\mathbf{c} = (1, -2)$, and $\mathbf{d} = (3, 1)$. The horizontal component of the vector $\mathbf{ab}$ is $1 - (-1) = 2$, and the vertical component of $\mathbf{ab}$ is $4 - 1 = 3$; thus, $\mathbf{ab} = (2, 3)$. Since the vector $\mathbf{cd}$ has a horizontal component of $3 - 1 = 2$, and a vertical component of $1 - (-2) = 3$, **cd** = (2, 3) also. Therefore, **ab** = **cd**; see Figure 6.

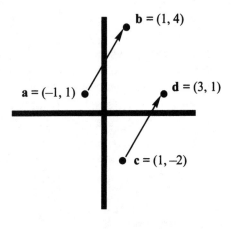

b = (1, 4)

a = (−1, 1)

d = (3, 1)

c = (1, −2)

■ Figure 6 ■

To **translate** a vector means to slide it so as to change its initial and terminal points but not its components. If the vector **ab** in Figure 6 were translated to begin at the point **c** = (1, −2), it would coincide with the vector **cd**. This is another way to say that **ab** = **cd**.

Example 4: Is the vector from **a** = (0, 2) to **b** = (3, 5) equivalent to the vector from **x** = (2, −4) to **y** = (5, 1)?

Since the vector **ab** equals (3, 3), but the vector **xy** equals (3, 5), these vectors are not equivalent. Alternatively, if the vector **ab** were translated to begin at the point **x**, its terminal point would then be $(x_1 + 3, x_2 + 3) = (2 + 3, -4 + 3) = (5, -1)$. This is not the point **y**; thus, **ab** ≠ **xy**. ■

Position vectors. If a vector has its initial point at the **origin**, the point $\mathbf{0} = (0, 0)$, it is called a **position vector**. If a position vector has $\mathbf{x} = (x_1, x_2)$ as its endpoint, then the components of the vector $\mathbf{0x}$ are $x_1 - 0 = x_1$ and $x_2 - 0 = x_2$; so $\mathbf{0x} = (x_1, x_2)$. If the origin is not explicitly written, then a position vector can be named by simply specifying its endpoint; thus, $\mathbf{x} = (x_1, x_2)$. Note that the position vector $\mathbf{x}$ with components x_1 and x_2 is denoted (x_1, x_2), just like the *point* $\mathbf{x}$ with *coordinates* x_1 and x_2. The context will make it clear which meaning is intended, but often the difference is irrelevant. Furthermore, since a position vector can be translated to begin at any other point in the plane without altering the vector (since translation leaves the components unchanged), even vectors that do not begin at the origin are named by a single letter.

Example 5: If the position vector $\mathbf{x} = (-4, 2)$ is translated so that its new initial point is $\mathbf{a} = (3, 1)$, find its new terminal point, $\mathbf{b}$.

If $\mathbf{b} = (b_1, b_2)$, then the components of the vector $\mathbf{ab}$ are $b_1 - 3$ and $b_2 - 1$. Since $\mathbf{ab} = \mathbf{x}$,

$$(b_1 - 3, b_2 - 1) = (-4, 2) \quad \Rightarrow \quad (b_1, b_2) = (-1, 3)$$

See Figure 7.

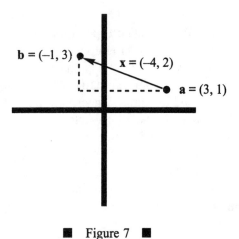

■ Figure 7 ■

Vector addition. The operations defined earlier on points (x_1, x_2) in $\mathbf{R}^2$ can be recast as operations on vectors in $\mathbf{R}^2$ (called **2-vectors**, because there are 2 components). These operations are called **vector addition** and **scalar multiplication**. The sum of two vectors $\mathbf{x}$ and $\mathbf{x}'$ is defined by the same rule that gave the sum of two points:

$$\mathbf{x} + \mathbf{x}' = (x_1,\ x_2) + (x_1',\ x_2') = (x_1 + x_1',\ x_2 + x_2')$$

Figure 8 depicts the sum of two vectors. Geometrically, one of the vectors ($\mathbf{x}'$, say) is translated so that its tail coincides with the tip of $\mathbf{x}$. The vector from the tail of $\mathbf{x}$ to the tip of the translated $\mathbf{x}'$ is the vector sum $\mathbf{x} + \mathbf{x}'$. This process is often referred to as adding vectors *tip-to-tail*.

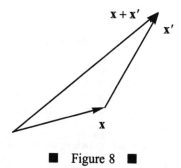

■ Figure 8 ■

Because the addition of real numbers is *commutative*, that is, because the order in which numbers are added is irrelevant, it follows that

$$(x_1 + x_1', \; x_2 + x_2') = (x_1' + x_1, \; x_2' + x_2)$$

This implies the addition of vectors is commutative also:

$$\mathbf{x} + \mathbf{x}' = \mathbf{x}' + \mathbf{x}$$

Thus, when adding $\mathbf{x}$ and $\mathbf{x}'$ geometrically, it doesn't matter whether $\mathbf{x}'$ is first translated to begin at the tip of $\mathbf{x}$ or $\mathbf{x}$ is translated to begin at the tip of $\mathbf{x}'$; the sum will be the same in either case.

Example 6: The sum of the vectors $\mathbf{x} = (1, 3)$ and $\mathbf{y} = (-2, 5)$ is

$$\mathbf{x} + \mathbf{y} = (1 + (-2), \; 3 + 5) = (-1, 8)$$

See Figure 9.

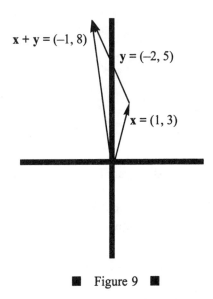

x + y = (−1, 8)

y = (−2, 5)

x = (1, 3)

■ Figure 9 ■

Example 7: Consider the position vector **a** = (1, 3). If **b** is the point (5, 4), find the vector **ab** and the vector sum **a** + **ab**. Provide a sketch.

Since **ab** has horizontal component 5 − 1 = 4 and vertical component 4 − 3 = 1, the vector **ab** equals (4, 1). So **a** + **ab** = (1, 3) + (4, 1) = (5, 4), which is the position vector **b**. Figure 10 clearly shows that **a** + **ab** = **b**.

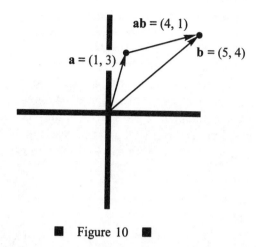

■ Figure 10 ■

Vector subtraction. The difference of two vectors is defined in precisely the same way as the difference of two points. For any two vectors **x** and **x'** in $\mathbf{R}^2$,

$$\mathbf{x} - \mathbf{x}' = (x_1, x_2) - (x_1', x_2') = (x_1 - x_1', x_2 - x_2')$$

With **x** and **x'** starting from the same point, **x − x'** is the vector that begins at the tip of **x'** and ends at the tip of **x**. This observation follows from the identity **x' + (x − x') = x** and the method of adding vectors geometrically. See Figure 11.

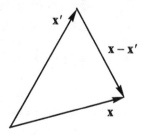

■ Figure 11 ■

In general, it is easy to see that

$$\mathbf{ab} = \mathbf{b} - \mathbf{a}$$

whether the letters on the right-hand side are interpreted as position vectors or as points. Figure 10 showed that $\mathbf{a} + \mathbf{ab} = \mathbf{b}$, which is equivalent to the statement $\mathbf{ab} = \mathbf{b} - \mathbf{a}$, where $\mathbf{a}$ and $\mathbf{b}$ are position vectors. Although this example dealt with a particular case, the identity $\mathbf{ab} = \mathbf{b} - \mathbf{a}$ holds in general.

Example 8: The vector $\mathbf{ab}$ from $\mathbf{a} = (4, -1)$ to $\mathbf{b} = (-2, 1)$ in Figure 12 is

$$\mathbf{ab} = \mathbf{b} - \mathbf{a} = (-2, 1) - (4, -1) = (-2 - 4, 1 + 1) = (-6, 2)$$

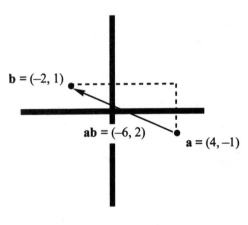

■ Figure 12 ■

Scalar multiplication. A vector $\mathbf{x}$ is multiplied by a scalar c by the rule

$$c\mathbf{x} = c(x_1, x_2) = (cx_1, cx_2)$$

If the scalar c is 0, then for any **x**, c**x** equals $(0, 0)$—the **zero vector**, denoted **0**. If c is positive, the vector c**x** points in the same direction as **x**, and it can be shown that its length is c times the length of **x**. However, if the scalar c is negative, then c**x** points in the direction exactly opposite to that of the original **x**, and the length of c**x** is $|c|$ times the length of **x**. Some examples are shown in Figure 13:

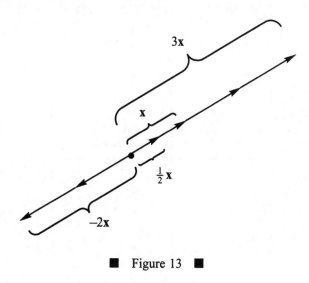

■ Figure 13 ■

Two vectors are said to be **parallel** if one is a positive scalar multiple of the other and **antiparallel** if one is a negative scalar multiple of the other. (Note: Some authors declare two vectors parallel if one is a scalar multiple—positive or negative—of the other.)

Standard basis vectors in $\mathbf{R}^2$. By invoking the definitions of vector addition and scalar multiplication, any vector $\mathbf{x} = (x_1, x_2)$ in $\mathbf{R}^2$ can be written in terms of the **standard basis vectors**

(1, 0) and (0, 1):

$$(x_1, x_2) = (x_1, 0) + (0, x_2) = x_1(1, 0) + x_2(0, 1)$$

The vector (1, 0) is denoted by **i** (or $\mathbf{e}_1$), and the vector (0, 1) is denoted by **j** (or $\mathbf{e}_2$). Using this notation, any vector **x** in $\mathbf{R}^2$ can be written in either of the two forms

$$\mathbf{x} = x_1\mathbf{i} + x_2\mathbf{j} \quad \text{or} \quad \mathbf{x} = x_1\mathbf{e}_1 + x_2\mathbf{e}_2$$

See Figure 14.

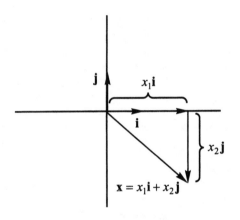

■ Figure 14 ■

Example 9: If **x** = 2**i** + 4**j** and **y** = **i** − 3**j**, determine (and provide a sketch of) the vectors $\frac{1}{2}$**x** and **x** + **y**.

Multiplying the vector **x** by the scalar $\frac{1}{2}$ yields

$$\tfrac{1}{2}\mathbf{x} = \tfrac{1}{2}(2\mathbf{i} + 4\mathbf{j}) = (\tfrac{1}{2} \cdot 2)\mathbf{i} + (\tfrac{1}{2} \cdot 4)\mathbf{j} = \mathbf{i} + 2\mathbf{j}$$

The sum of the vectors **x** and **y** is

$$\mathbf{x} + \mathbf{y} = (2\mathbf{i} + 4\mathbf{j}) + (\mathbf{i} - 3\mathbf{j}) = (2 + 1)\mathbf{i} + (4 - 3)\mathbf{j} = 3\mathbf{i} + \mathbf{j}$$

These vectors are shown (together with $\mathbf{x}$ and $\mathbf{y}$) in Figure 15.

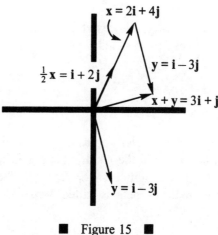

■ Figure 15 ■

Example 10: Find the scalar coefficients k_1 and k_2 such that
$$k_1(1, -3) + k_2(-1, 2) = (-1, -2)$$

The given equation can be rewritten as follows:
$$(k_1 - k_2, -3k_1 + 2k_2) = (-1, -2)$$

This implies that both of the following equations must be satisfied:

$$\begin{aligned} k_1 - k_2 &= -1 \\ -3k_1 + 2k_2 &= -2 \end{aligned} \quad (*)$$

Multiplying the first equation by 3 then adding the result to the second equation yields

$$3k_1 - 3k_2 = -3$$
$$\underline{-3k_1 + 2k_2 = -2}$$
$$-k_2 = -5$$

Thus, $k_2 = 5$. Substituting this result back into either of the equations in (*) gives $k_1 = 4$. ■

The Space $\mathbf{R}^3$

If *three* mutually perpendicular copies of the real line intersect at their origins, any point in the resulting space is specified by an ordered *triple* of real numbers (x_1, x_2, x_3). The set of all ordered triples of real numbers is called **3-space**, denoted $\mathbf{R}^3$ ("R three"). See Figure 16.

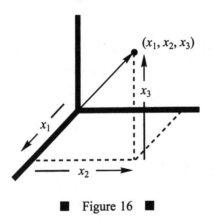

■ Figure 16 ■

The operations of addition and scalar multiplication defined on $\mathbf{R}^2$ carry over to $\mathbf{R}^3$:

$$(x_1, x_2, x_3) + (x_1', x_2', x_3') = (x_1 + x_1', x_2 + x_2', x_3 + x_3')$$
$$c(x_1, x_2, x_3) = (cx_1, cx_2, cx_3)$$

Vectors in $\mathbf{R}^3$ are called **3-vectors** (because there are 3 components), and the geometric descriptions of addition and scalar multiplication given for 2-vectors also carry over to 3-vectors.

Example 11: If $\mathbf{x} = (3, 0, 4)$ and $\mathbf{y} = (2, 1, -1)$, then
$$3\mathbf{x} - 2\mathbf{y} = 3(3, 0, 4) - 2(2, 1, -1)$$
$$= (9, 0, 12) - (4, 2, -2) = (5, -2, 14) \quad \blacksquare$$

Standard basis vectors in $\mathbf{R}^3$. Since for any vector $\mathbf{x} = (x_1, x_2, x_3)$ in $\mathbf{R}^3$,

$$(x_1, x_2, x_3) = (x_1, 0, 0) + (0, x_2, 0) + (0, 0, x_3)$$
$$= x_1(1, 0, 0) + x_2(0, 1, 0) + x_3(0, 0, 1)$$

the standard basis vectors in $\mathbf{R}^3$ are

$$\mathbf{i} = \mathbf{e}_1 = (1, 0, 0), \quad \mathbf{j} = \mathbf{e}_2 = (0, 1, 0), \quad \text{and} \quad \mathbf{k} = \mathbf{e}_3 = (0, 0, 1)$$

Any vector $\mathbf{x}$ in $\mathbf{R}^3$ may therefore be written as

$$\mathbf{x} = x_1\mathbf{i} + x_2\mathbf{j} + x_3\mathbf{k} \quad \text{or} \quad \mathbf{x} = x_1\mathbf{e}_1 + x_2\mathbf{e}_2 + x_3\mathbf{e}_3$$

See Figure 17.

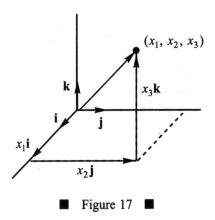

■ Figure 17 ■

Example 12: What vector must be added to **a** = (1, 3, 1) to yield **b** = (3, 1, 5)?

Let **c** be the required vector; then **a** + **c** = **b**. Therefore,

$$\mathbf{c} = \mathbf{b} - \mathbf{a} = (3, 1, 5) - (1, 3, 1) = (2, -2, 4)$$

Note that **c** is the vector **ab**; see Figure 18.

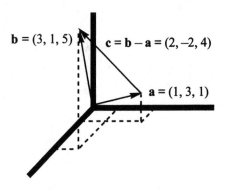

■ Figure 18 ■

The cross product. So far, you have seen how two vectors can be added (or subtracted) and how a vector is multiplied by a scalar. Is it possible to somehow "multiply" two vectors? One way to define the product of two vectors—which is done only with vectors in $\mathbf{R}^3$—is to form their *cross product*. Let $\mathbf{x} = (x_1, x_2, x_3)$ and $\mathbf{y} = (y_1, y_2, y_3)$ be two vectors in $\mathbf{R}^3$. The **cross product** (or **vector product**) of $\mathbf{x}$ and $\mathbf{y}$ is defined as follows:

$$\mathbf{x} \times \mathbf{y} = (x_1, x_2, x_3) \times (y_1, y_2, y_3)$$

$$= (x_2 y_3 - x_3 y_2, \ x_3 y_1 - x_1 y_3, \ x_1 y_2 - x_2 y_1)$$

The cross product of two vectors is a vector, and perhaps the most important characteristic of this vector product is that *it is perpendicular to both factors*. (This will be demonstrated when the dot product is introduced.) That is, the vector $\mathbf{x} \times \mathbf{y}$ will be perpendicular to both $\mathbf{x}$ and $\mathbf{y}$; see Figure 19. [There is an ambiguity here: the plane in Figure 19, which contains the vectors $\mathbf{x}$ and $\mathbf{y}$, has two perpendicular directions: "up" and "down." Which one does the cross product choose? The answer is given by the *right-hand rule*: Place the wrist of your right hand at the common initial point of $\mathbf{x}$ and $\mathbf{y}$, with your fingers pointing along $\mathbf{x}$; as you curl your fingers toward $\mathbf{y}$, your thumb will point in the direction of $\mathbf{x} \times \mathbf{y}$. This shows that the cross product is anticommutative: $\mathbf{y} \times \mathbf{x} = -(\mathbf{x} \times \mathbf{y})$.]

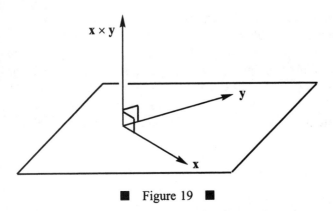

■ Figure 19 ■

The **length** of a vector $\mathbf{x} = (x_1, x_2, x_3)$ in $\mathbf{R}^3$, which is denoted $\|\mathbf{x}\|$, is given by the equation

$$\|\mathbf{x}\| = \sqrt{(x_1)^2 + (x_2)^2 + (x_3)^2}$$

a result which follows from the Pythagorean Theorem (see the discussion preceding Figure 22 below). While the direction of the cross product of $\mathbf{x}$ and $\mathbf{y}$ is determined by orthogonality and the right-hand rule, the **magnitude** (that is, the length) of $\mathbf{x} \times \mathbf{y}$ is equal to the area of the parallelogram spanned by the vectors $\mathbf{x}$ and $\mathbf{y}$.

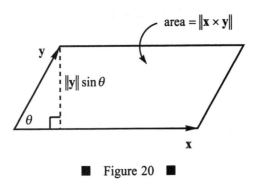

■ Figure 20 ■

Since the area of the parallelogram in Figure 20 is

$$\text{area} = \text{base} \cdot \text{height} = \|\mathbf{x}\| \cdot \|\mathbf{y}\| \sin \theta$$

the following equation holds:

$$\boxed{\|\mathbf{x} \times \mathbf{y}\| = \|\mathbf{x}\| \cdot \|\mathbf{y}\| \sin \theta}$$

where θ is the angle between $\mathbf{x}$ and $\mathbf{y}$.

Example 13: Let $\mathbf{x} = (2, 3, 0)$ and $\mathbf{y} = (-1, 1, 4)$ be position vectors in $\mathbf{R}^3$. Compute the area of the triangle whose vertices are the origin and the endpoints of $\mathbf{x}$ and $\mathbf{y}$ and determine the angle between the vectors $\mathbf{x}$ and $\mathbf{y}$.

Since the area of the triangle is half the area of the parallelogram spanned by $\mathbf{x}$ and $\mathbf{y}$,

$$\begin{aligned}
\text{area of } \Delta &= \tfrac{1}{2} \|\mathbf{x} \times \mathbf{y}\| \\
&= \tfrac{1}{2} \|(x_2 y_3 - x_3 y_2, \ x_3 y_1 - x_1 y_3, \ x_1 y_2 - x_2 y_1)\| \\
&= \tfrac{1}{2} \|(3 \cdot 4 - 0 \cdot 1, \ 0 \cdot (-1) - 2 \cdot 4, \ 2 \cdot 1 - 3 \cdot (-1)\| \\
&= \tfrac{1}{2} \|(12, \ -8, \ 5)\| \\
&= \tfrac{1}{2} \sqrt{12^2 + (-8)^2 + 5^2} \\
&= \tfrac{1}{2} \sqrt{233}
\end{aligned}$$

Now, since $\|\mathbf{x} \times \mathbf{y}\| = \|\mathbf{x}\| \cdot \|\mathbf{y}\| \sin \theta$, the angle between $\mathbf{x}$ and $\mathbf{y}$ is given by

$$\sin\theta = \frac{\|\mathbf{x} \times \mathbf{y}\|}{\|\mathbf{x}\|\,\|\mathbf{y}\|} = \frac{\sqrt{233}}{\sqrt{2^2 + 3^2 + 0^2}\cdot\sqrt{(-1)^2 + 1^2 + 4^2}}$$

$$= \frac{\sqrt{233}}{\sqrt{13}\cdot\sqrt{18}}$$

$$= \sqrt{\frac{233}{234}}$$

Therefore, $\theta = \sin^{-1}\sqrt{233/234}$. ∎

The Space $\mathbf{R}^n$

By analogy with the preceding constructions ($\mathbf{R}^2$ and $\mathbf{R}^3$), you can consider the collection of all ordered **n-tuples** of real numbers $(x_1, x_2, \ldots, x_n)$ with the analogous operations of addition and scalar multiplication. This is called **n-space** (denoted $\mathbf{R}^n$), and vectors in $\mathbf{R}^n$ are called **n-vectors**. The standard basis vectors in $\mathbf{R}^n$ are

$$\mathbf{e}_1 = (1, 0, 0, \ldots, 0),$$
$$\mathbf{e}_2 = (0, 1, 0, \ldots, 0), \ldots,$$
$$\mathbf{e}_n = (0, 0, 0, \ldots, 1)$$

where $\mathbf{e}_k$ has a 1 in the kth place and zeros elsewhere. All the figures above depicted points and vectors in $\mathbf{R}^2$ and $\mathbf{R}^3$. Although it is not possible to draw such diagrams to illustrate geometric figures in $\mathbf{R}^n$ if $n > 3$, it *is* possible to deal with them *algebraically*, and therein lies the real power of the algebraic machinery.

Example 14: Consider the vectors $\mathbf{a} = (1, 2, 0, -3)$, $\mathbf{b} = (0, 1, -4, 2)$, and $\mathbf{c} = (5, -1, -1, 1)$ in $\mathbf{R}^4$. Determine the vector $2\mathbf{a} - \mathbf{b} + \mathbf{c}$.

Extend the definitions of scalar multiplication and vector addition in the natural way to vectors in $\mathbf{R}^4$ to compute

$$
\begin{aligned}
2\mathbf{a} - \mathbf{b} + \mathbf{c} &= 2(1, 2, 0, -3) - (0, 1, -4, 2) + (5, -1, -1, 1) \\
&= (2, 4, 0, -6) - (0, 1, -4, 2) + (5, -1, -1, 1) \\
&= (2 - 0 + 5, \ 4 - 1 - 1, \ 0 + 4 - 1, \ -6 - 2 + 1) \\
&= (7, 2, 3, -7) \quad \blacksquare
\end{aligned}
$$

Example 15: Determine the sum of the standard basis vectors $\mathbf{e}_1$, $\mathbf{e}_3$, and $\mathbf{e}_4$ in $\mathbf{R}^5$.

All vectors in $\mathbf{R}^5$ have five components. Four of the components in each of the standard basis vectors in $\mathbf{R}^5$ are zero, and one component—the first in $\mathbf{e}_1$, the third in $\mathbf{e}_3$, and the fourth in $\mathbf{e}_4$—has the value 1. Therefore,

$$
\begin{aligned}
\mathbf{e}_1 + \mathbf{e}_3 + \mathbf{e}_4 &= (1, 0, 0, 0, 0) + (0, 0, 1, 0, 0) + (0, 0, 0, 1, 0) \\
&= (1, 0, 1, 1, 0) \quad \blacksquare
\end{aligned}
$$

The norm of a vector. The **length** (or **Euclidean norm**) of a vector $\mathbf{x}$ is denoted $\|\mathbf{x}\|$, and for a vector $\mathbf{x} = (x_1, x_2)$ in $\mathbf{R}^2$, $\|\mathbf{x}\|$ is easy to compute (see Figure 21) by applying the Pythagorean Theorem:

$$
\|\mathbf{x}\| = \sqrt{(x_1)^2 + (x_2)^2}
$$

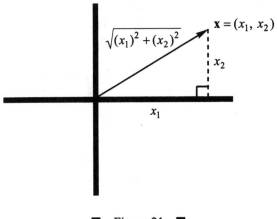

■ Figure 21 ■

The expression for the length of a vector $\mathbf{x} = (x_1, x_2, x_3)$ in $\mathbf{R}^3$ follows from two applications of the Pythagorean Theorem, as illustrated in Figure 22:

$$\|\mathbf{x}\| = \sqrt{(x_1)^2 + (x_2)^2 + (x_3)^2}$$

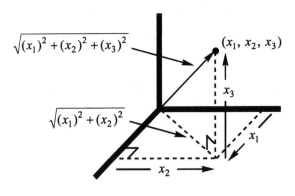

■ Figure 22 ■

In general, the norm of a vector $\mathbf{x} = (x_1, x_2, x_3, \ldots, x_n)$ in $\mathbf{R}^n$ is given by the equation

$$\|\mathbf{x}\| = \sqrt{(x_1)^2 + (x_2)^2 + \cdots + (x_n)^2}$$

Example 16: The length of the vector $\mathbf{x} = (3, 1, -5, 1)$ in $\mathbf{R}^4$ is

$$\|\mathbf{x}\| = \sqrt{3^2 + 1^2 + (-5)^2 + 1^2} = \sqrt{36} = 6 \quad \blacksquare$$

Example 17: Let $\mathbf{x}$ be a vector in $\mathbf{R}^n$. If c is a scalar, how does the norm of $c\mathbf{x}$ compare to the norm of $\mathbf{x}$?

If $\mathbf{x} = (x_1, x_2, \ldots, x_n)$, then $c\mathbf{x} = (cx_1, cx_2, \ldots, cx_n)$. Therefore,

$$\begin{aligned}
\|c\mathbf{x}\| &= \sqrt{(cx_1)^2 + (cx_2)^2 + \cdots + (cx_n)^2} \\
&= \sqrt{c^2\left[(x_1)^2 + (x_2)^2 + \cdots + (x_n)^2\right]} \\
&= \sqrt{c^2} \cdot \sqrt{(x_1)^2 + (x_2)^2 + \cdots + (x_n)^2} \\
&= |c| \cdot \|\mathbf{x}\|
\end{aligned}$$

Thus, multiplying a vector by a scalar c multiplies its norm by $|c|$. Note that this is consistent with the geometric description given earlier for scalar multiplication. $\blacksquare$

Distance between two points. The distance between two points $\mathbf{x}$ and $\mathbf{y}$ in $\mathbf{R}^n$—a quantity denoted by $d(\mathbf{x}, \mathbf{y})$—is defined to be the length of the vector $\mathbf{xy}$:

$$d(\mathbf{x}, \mathbf{y}) = \|\mathbf{xy}\|$$

Example 18: What is the distance between the points **p** = (3, 1, 4) and **q** = (1, 3, 2)?

Since **pq** = **q** − **p** = (1, 3, 2) − (3, 1, 4) = (−2, 2, −2), the distance between the points **p** and **q** is

$$d(\mathbf{p},\ \mathbf{q}) = \|\mathbf{pq}\| = \|(-2,\ 2,\ -2)\| = \sqrt{(-2)^2 + 2^2 + (-2)^2} = 2\sqrt{3} \quad \blacksquare$$

Unit vectors. Any vector whose length is 1 is called a **unit vector**. Let **x** be a given nonzero vector and consider the scalar multiple **x**/‖**x**‖. (The zero vector must be excluded from consideration here, for if **x** were **0**, then ‖**x**‖ would be 0, and the expression **x**/‖**x**‖ would be undefined.) Applying the result of Example 17 (with $c = 1/\|\mathbf{x}\|$), the norm of the vector **x**/‖**x**‖ is

$$\left\| \frac{1}{\|\mathbf{x}\|} \mathbf{x} \right\| = \frac{1}{\|\mathbf{x}\|} \|\mathbf{x}\| = 1$$

Thus, for any nonzero vector **x**,

$$\frac{\mathbf{x}}{\|\mathbf{x}\|}$$

is a unit vector. This vector is denoted $\hat{\mathbf{x}}$ ("x hat") and represents the unit vector in the direction of **x**. (Indeed, one can go further and actually call $\hat{\mathbf{x}}$ the **direction of x**.) Note in particular that all the standard basis vectors are unit vectors; they are sometimes written as

$$\hat{\mathbf{i}},\ \hat{\mathbf{j}},\ \text{etc. (or } \hat{\mathbf{e}}_1,\ \hat{\mathbf{e}}_2,\ \text{etc.)}$$

to emphasize this fact.

Example 19: Find the vector **y** in $\mathbf{R}^2$ whose length is 10 and which has the same direction as $\mathbf{x} = 3\mathbf{i} + 4\mathbf{j}$.

The idea is simple: Find the unit vector in the same direction as $3\mathbf{i} + 4\mathbf{j}$, and then multiply this unit vector by 10. The unit vector in the direction of **x** is

$$\hat{\mathbf{x}} = \frac{\mathbf{x}}{\|\mathbf{x}\|} = \frac{3\mathbf{i}+4\mathbf{j}}{\sqrt{3^2+4^2}} = \frac{3\mathbf{i}+4\mathbf{j}}{5} = \tfrac{3}{5}\mathbf{i} + \tfrac{4}{5}\mathbf{j}$$

Therefore,

$$\mathbf{y} = 10\hat{\mathbf{x}} = 10\left(\tfrac{3}{5}\mathbf{i} + \tfrac{4}{5}\mathbf{j}\right) = 6\mathbf{i} + 8\mathbf{j} \qquad \blacksquare$$

The dot product. One way to multiply two vectors—if they lie in $\mathbf{R}^3$—is to form their cross product. Another way to form the product of two vectors—from the same space $\mathbf{R}^n$, for *any* n—is as follows. For any two n-vectors $\mathbf{x} = (x_1, x_2, \ldots, x_n)$ and $\mathbf{y} = (y_1, y_2, \ldots, y_n)$, their **dot product** (or **Euclidean inner product**) is defined by the equation

$$\mathbf{x} \cdot \mathbf{y} = x_1 y_1 + x_2 y_2 + \cdots + x_n y_n$$

(The symbol $\mathbf{x} \cdot \mathbf{y}$ is read "x dot y.") Note carefully that, unlike the cross product, the dot product of two vectors is a *scalar*. For this reason, the dot product is also called the **scalar product**. It can be easily shown that the dot product on $\mathbf{R}^n$ satisfies the following useful identities:

Homogeneity:	$(c\mathbf{x}) \cdot \mathbf{y} = \mathbf{x} \cdot (c\mathbf{y}) = c(\mathbf{x} \cdot \mathbf{y})$
Commutative property:	$\mathbf{x} \cdot \mathbf{y} = \mathbf{y} \cdot \mathbf{x}$
Distributive property:	$\mathbf{x} \cdot (\mathbf{y} \pm \mathbf{z}) = \mathbf{x} \cdot \mathbf{y} \pm \mathbf{x} \cdot \mathbf{z}$

Example 20: What is the dot product of the vectors $\mathbf{x} = (-1, 0, 4)$ and $\mathbf{y} = (3, 6, 2)$ in $\mathbf{R}^3$?

By the commutative property, it doesn't matter whether the product is taken to be $\mathbf{x} \cdot \mathbf{y}$ or $\mathbf{y} \cdot \mathbf{x}$; the result is the same in either case. Applying the definition yields

$$\mathbf{x} \cdot \mathbf{y} = (-1)(3) + (0)(6) + (4)(2) = -3 + 0 + 8 = 5 \qquad \blacksquare$$

The dot product of a vector $\mathbf{x} = (x_1, x_2, \ldots, x_n)$ with itself is

$$\mathbf{x} \cdot \mathbf{x} = x_1 x_1 + x_2 x_2 + \cdots + x_n x_n = (x_1)^2 + (x_2)^2 + \cdots + (x_n)^2$$

Notice that the right-hand side of this equation is also the expression for $\|\mathbf{x}\|^2$:

$$\|\mathbf{x}\|^2 = (x_1)^2 + (x_2)^2 + \cdots + (x_n)^2$$

Therefore, for any vector $\mathbf{x}$,

$$\boxed{\|\mathbf{x}\|^2 = \mathbf{x} \cdot \mathbf{x}}$$

This identity is put to use as follows. Since $\|\mathbf{a}\|^2 = \mathbf{a} \cdot \mathbf{a}$, the distributive and commutative properties of the dot product imply that for any vectors $\mathbf{x}$ and $\mathbf{y}$ in $\mathbf{R}^n$,

$$\|\mathbf{x}+\mathbf{y}\|^2 = (\mathbf{x}+\mathbf{y})\cdot(\mathbf{x}+\mathbf{y})$$
$$= (\mathbf{x}+\mathbf{y})\cdot\mathbf{x}+(\mathbf{x}+\mathbf{y})\cdot\mathbf{y}$$
$$= \mathbf{x}\cdot(\mathbf{x}+\mathbf{y})+\mathbf{y}\cdot(\mathbf{x}+\mathbf{y})$$
$$= (\mathbf{x}\cdot\mathbf{x}+\mathbf{x}\cdot\mathbf{y})+(\mathbf{y}\cdot\mathbf{x}+\mathbf{y}\cdot\mathbf{y})$$
$$= \mathbf{x}\cdot\mathbf{x}+\mathbf{x}\cdot\mathbf{y}+\mathbf{x}\cdot\mathbf{y}+\mathbf{y}\cdot\mathbf{y}$$
$$= \mathbf{x}\cdot\mathbf{x}+2\mathbf{x}\cdot\mathbf{y}+\mathbf{y}\cdot\mathbf{y}$$

Thus, $\qquad \|\mathbf{x}+\mathbf{y}\|^2 = \|\mathbf{x}\|^2 + 2\mathbf{x}\cdot\mathbf{y} + \|\mathbf{y}\|^2 \quad$ (*)

Now, if $\mathbf{x}\perp\mathbf{y}$, then by Figure 23, the Pythagorean Theorem would say

$$\|\mathbf{x}+\mathbf{y}\|^2 = \|\mathbf{x}\|^2 + \|\mathbf{y}\|^2 \qquad (**)$$

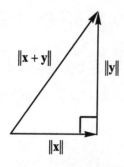

■ Figure 23 ■

Therefore, if $\mathbf{x}\perp\mathbf{y}$, equations (*) and (**) imply

$$\|\mathbf{x}\|^2 + \|\mathbf{y}\|^2 = \|\mathbf{x}\|^2 + 2\mathbf{x}\cdot\mathbf{y} + \|\mathbf{y}\|^2$$

which simplifies to the simple statement $\mathbf{x}\cdot\mathbf{y}=0$. Since this argument is reversible (assuming that it is agreed that the zero vector is orthogonal to *every* vector), the following fact has been established:

$$\boxed{\;\mathbf{x}\perp\mathbf{y}\quad\text{if and only if}\quad\mathbf{x}\cdot\mathbf{y}=0\;}$$

This says that two vectors are **orthogonal**—that is, perpendicular—if and only if their dot product is zero.

Example 21: Use the dot product to verify that the cross product of the vectors $\mathbf{x} = (2, 3, 0)$ and $\mathbf{y} = (-1, 1, 4)$ from Example 13 is orthogonal to both $\mathbf{x}$ and $\mathbf{y}$; then show that $\mathbf{x}\times\mathbf{y}$ is orthogonal to both $\mathbf{x}$ and $\mathbf{y}$ for *any* vectors $\mathbf{x}$ and $\mathbf{y}$ in $\mathbf{R}^3$.

In Example 13, it was determined that $\mathbf{x}\times\mathbf{y} = (12, -8, 5)$. The criterion for orthogonality is the vanishing of the dot product. Since both

$$(\mathbf{x}\times\mathbf{y})\cdot\mathbf{x} = (12, -8, 5)\cdot(2, 3, 0) = 12\cdot 2 - 8\cdot 3 + 5\cdot 0 = 0$$

and

$$(\mathbf{x}\times\mathbf{y})\cdot\mathbf{y} = (12, -8, 5)\cdot(-1, 1, 4) = 12\cdot(-1) - 8\cdot 1 + 5\cdot 4 = 0$$

the vector $\mathbf{x}\times\mathbf{y}$ is indeed orthogonal to $\mathbf{x}$ and to $\mathbf{y}$. In general,

$$
\begin{aligned}
(\mathbf{x}\times\mathbf{y})\cdot\mathbf{y} &= \mathbf{y}\cdot(\mathbf{x}\times\mathbf{y})\\
&= (y_1,\, y_2,\, y_3)\cdot(x_2y_3 - x_3y_2,\; x_3y_1 - x_1y_3,\; x_1y_2 - x_2y_1)\\
&= y_1(x_2y_3 - x_3y_2) + y_2(x_3y_1 - x_1y_3) + y_3(x_1y_2 - x_2y_1)\\
&= (y_1x_2y_3 - y_3x_2y_1) + (-y_1x_3y_2 + y_2x_3y_1)\\
&\qquad\qquad\qquad\qquad + (-y_2x_1y_3 + y_3x_1y_2)\\
&= 0 + 0 + 0\\
&= 0
\end{aligned}
$$

and a similar calculation shows that $(\mathbf{x}\times\mathbf{y})\cdot\mathbf{x} = 0$ also. ∎

The triangle inequality. From elementary geometry, you know that the sum of the lengths of any two sides of a triangle must be greater than the length of the third side. That is, if A, B, and C are the vertices of a triangle, then

$$AC < AB + BC$$

This just says that the direct journey from A to C (along side AC) is shorter than the path from A to B and then to C; see Figure 24. This is called the **triangle inequality**.

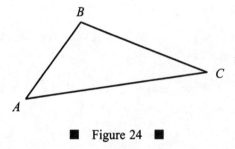

■ Figure 24 ■

The triangle inequality can be generalized to vectors in $\mathbf{R}^n$. If $\mathbf{x}$ and $\mathbf{y}$ are any two n-vectors, then

$$\|\mathbf{x} + \mathbf{y}\| \leq \|\mathbf{x}\| + \|\mathbf{y}\|$$

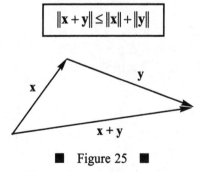

■ Figure 25 ■

Figure 25 shows that this statement can be interpreted in the same way as the elementary geometric fact about the lengths of the sides of a triangle. [One notable difference, however, is that if **x** and **y** happen to be parallel (that is, if **y** is a positive scalar times **x**) or if either **x** or **y** is the zero vector, then $\|\mathbf{x} + \mathbf{y}\| = \|\mathbf{x}\| + \|\mathbf{y}\|$. The generalized triangle inequality must take these degenerate cases into account (hence the *weak* inequality, $\leq$), whereas the triangle inequality from elementary geometry does not (and hence the *strong* (or *strict*) inequality, $<$).]

Example 22: Verify the triangle inequality for the vectors **x** = (–1, 0, 4) and **y** = (3, 6, 2) from Example 20.

The sum of these vectors is **x** + **y** = (2, 6, 6), and the lengths of the vectors **x**, **y**, and **x** + **y** are

$$\|\mathbf{x}\| = \sqrt{(-1)^2 + 0^2 + 4^2} = \sqrt{17}$$

$$\|\mathbf{y}\| = \sqrt{3^2 + 6^2 + 2^2} = \sqrt{49} = 7$$

$$\|\mathbf{x} + \mathbf{y}\| = \sqrt{2^2 + 6^2 + 6^2} = \sqrt{76}$$

With these lengths, the triangle inequality, $\|\mathbf{x} + \mathbf{y}\| \leq \|\mathbf{x}\| + \|\mathbf{y}\|$, becomes $\sqrt{76} \leq \sqrt{17} + 7$, which is certainly true, since the left-hand side is less than 9, while the right-hand side is greater than 4 + 7 = 11. ∎

The Cauchy-Schwarz inequality. One of the most important inequalities in mathematics is known as the **Cauchy-Schwarz inequality**. For $\mathbf{R}^n$ equipped with an inner product, this inequality states

$$\boxed{\;|\mathbf{x}\cdot\mathbf{y}| \le \|\mathbf{x}\|\,\|\mathbf{y}\|\;}$$

which says the absolute value of the dot product of two vectors is never greater than the product of their norms. Because of this inequality, it must be true that for any two nonzero vectors $\mathbf{x}$ and $\mathbf{y}$,

$$\frac{|\mathbf{x}\cdot\mathbf{y}|}{\|\mathbf{x}\|\,\|\mathbf{y}\|} \le 1$$

Since both $\|\mathbf{x}\|$ and $\|\mathbf{y}\|$ are positive, the absolute value signs can be repositioned:

$$\left|\frac{\mathbf{x}\cdot\mathbf{y}}{\|\mathbf{x}\|\,\|\mathbf{y}\|}\right| \le 1$$

a statement which now directly implies

$$-1 \le \frac{\mathbf{x}\cdot\mathbf{y}}{\|\mathbf{x}\|\,\|\mathbf{y}\|} \le 1$$

This final inequality says that there is precisely one value of θ between 0 and π (inclusive) such that

$$\frac{\mathbf{x}\cdot\mathbf{y}}{\|\mathbf{x}\|\,\|\mathbf{y}\|} = \cos\theta$$

This θ is called the **angle between the vectors x and y**; geometrically, it is the smaller angle between them. (Note: No angle θ is defined if either $\mathbf{x}$ or $\mathbf{y}$ is the zero vector.) To verify that this θ is indeed the geometric angle between $\mathbf{x}$ and $\mathbf{y}$, consider Figure 26, where it is assumed that the angle between $\mathbf{x}$ and $\mathbf{y}$ is acute.

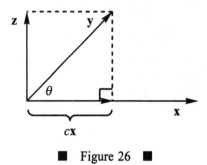

■ Figure 26 ■

The vector **z** is orthogonal to **x**, and the figure shows that **y** is the sum of **z** and a positive scalar multiple, $c\mathbf{x}$, of **x**:

$$c\mathbf{x} + \mathbf{z} = \mathbf{y}$$

Taking the dot product of both sides of this equation with **x** yields

$$c\mathbf{x} + \mathbf{z} = \mathbf{y}$$
$$\mathbf{x} \cdot (c\mathbf{x} + \mathbf{z}) = \mathbf{x} \cdot \mathbf{y}$$
$$c(\mathbf{x} \cdot \mathbf{x}) + \mathbf{x} \cdot \mathbf{z} = \mathbf{x} \cdot \mathbf{y}$$

Since **x** and **z** are orthogonal, the dot product $\mathbf{x} \cdot \mathbf{z}$ is 0. This reduces the equation above to

$$c = \frac{\mathbf{x} \cdot \mathbf{y}}{\mathbf{x} \cdot \mathbf{x}} \quad (*)$$

But Figure 26 and the definition of $\cos\theta$ indicate that

$$\|c\mathbf{x}\| = \|\mathbf{y}\|\cos\theta$$

Now, since c is positive, $\|c\mathbf{x}\| = |c|\,\|\mathbf{x}\| = c\,\|\mathbf{x}\|$, so this equation becomes

$$c = \frac{\|\mathbf{y}\| \cos\theta}{\|\mathbf{x}\|} \quad (**)$$

Equations (*) and (**), together with the identity $\mathbf{x} \cdot \mathbf{x} = \|\mathbf{x}\|^2$, then imply

$$\frac{\mathbf{x} \cdot \mathbf{y}}{\|\mathbf{x}\|^2} = \frac{\|\mathbf{y}\| \cos\theta}{\|\mathbf{x}\|}$$

which becomes

$$\mathbf{x} \cdot \mathbf{y} = \|\mathbf{x}\| \|\mathbf{y}\| \cos\theta$$

This proof can be extended to the case where the angle between $\mathbf{x}$ and $\mathbf{y}$ is obtuse, thus validating the following alternate—but entirely equivalent—definition of the dot product:

$$\boxed{\mathbf{x} \cdot \mathbf{y} = \|\mathbf{x}\| \|\mathbf{y}\| \cos\theta}$$

Note that this equation is consistent with the observation $\mathbf{x} \perp \mathbf{y} \Rightarrow \mathbf{x} \cdot \mathbf{y} = 0$, since $\theta = \pi/2$ implies $\cos\theta = 0$.

Example 23: Use the dot product to determine the angle between the vectors $\mathbf{x} = (2, 3, 0)$ and $\mathbf{y} = (-1, 1, 4)$ from Example 13.

From the boxed equation directly above,

$$\cos\theta = \frac{\mathbf{x}\cdot\mathbf{y}}{\|\mathbf{x}\|\|\mathbf{y}\|}$$

$$= \frac{2\cdot(-1)+3\cdot 1+0\cdot 4}{\sqrt{2^2+3^2+0^2}\sqrt{(-1)^2+1^2+4^2}}$$

$$= \frac{1}{\sqrt{234}}$$

Therefore, $\theta = \cos^{-1}\sqrt{1/234}$.

[Technical note: In Example 13, it was determined that $\theta = \sin^{-1}\sqrt{233/234}$. Although this is consistent with the present calculation (because $\cos^2\theta+\sin^2\theta$ must always equal 1 for any θ, and this is certainly true here), it is better to use the dot product than the cross product to determine the angle between two vectors in $\mathbf{R}^3$. Why? The statement $\sin\theta = \sqrt{233/234}$ implies that θ is either 86.25° or 93.75°, and without further investigation, it is difficult to say which is correct. Even a picture may not help here; the angle is so close to 90° that your drawing will probably not be accurate enough to tell the difference. But the statement $\cos\theta = \sqrt{1/234}$ says that θ is definitely 86.25°, with no ambiguity. Within the range between 0 and 180°, the sine function is entirely positive and cannot differentiate between an acute angle and its supplement. However, the cosine function is positive for acute angles and negative for obtuse angles, so it can—immediately—differentiate between an acute angle and its supplement.] ∎

Orthogonal projections. Consider two nonzero vectors **x** and **y** emanating from the origin in $\mathbf{R}^n$. Dropping a perpendicular from the tip of **x** to the line containing **y** gives the (**orthogonal**) **projection of x onto y**. This vector is denoted **proj$_\mathbf{y}$x**.

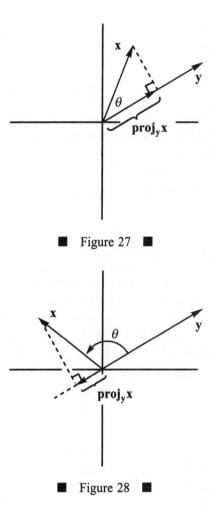

■ Figure 27 ■

■ Figure 28 ■

If $\theta < \pi/2$ (Figure 27), then the **component of x along y**, a positive scalar denoted $\text{comp}_y\mathbf{x}$, is equal to the norm of the (vector) projection of **x** onto **y**. If $\theta > \pi/2$ (Figure 28), then the component of **x** along **y** is a negative scalar, equal to the negative of the norm of the projection of **x** onto **y**. (And if

$\theta = \pi/2$, then $\text{comp}_y \mathbf{x} = 0$, since the orthogonal projection of $\mathbf{x}$ onto $\mathbf{y}$ is the zero vector.) In any case, the following equation holds:

$$\text{comp}_y \mathbf{x} = \|\mathbf{x}\| \cos \theta$$

where θ is the angle between $\mathbf{x}$ and $\mathbf{y}$. Now, since $\mathbf{x} \cdot \mathbf{y} = \|\mathbf{x}\| \|\mathbf{y}\| \cos\theta$, this equation for the component of $\mathbf{x}$ along $\mathbf{y}$ can be rewritten as

$$\text{comp}_y \mathbf{x} = \|\mathbf{x}\| \frac{\mathbf{x} \cdot \mathbf{y}}{\|\mathbf{x}\| \|\mathbf{y}\|} = \frac{\mathbf{x} \cdot \mathbf{y}}{\|\mathbf{y}\|}$$

The vector projection of $\mathbf{x}$ onto $\mathbf{y}$ is equal to this scalar times the unit vector in the direction of $\mathbf{y}$:

$$\boxed{\mathbf{proj}_y \mathbf{x} = (\text{comp}_y \mathbf{x})\hat{\mathbf{y}}}$$

$$= \frac{\mathbf{x} \cdot \mathbf{y}}{\|\mathbf{y}\|} \frac{\mathbf{y}}{\|\mathbf{y}\|}$$

Or, since $\|\mathbf{y}\|^2 = \mathbf{y} \cdot \mathbf{y}$,

$$\mathbf{proj}_y \mathbf{x} = \frac{\mathbf{x} \cdot \mathbf{y}}{\mathbf{y} \cdot \mathbf{y}} \mathbf{y}$$

Example 24: Find the projection of $\mathbf{x} = (2, 2, 4)$ onto the vector $\mathbf{y} = (2, 6, 3)$.

If θ is the angle between $\mathbf{x}$ and $\mathbf{y}$, then the component of $\mathbf{x}$ along $\mathbf{y}$ is given by

$$\text{comp}_y \, \mathbf{x} = \|\mathbf{x}\| \cos \theta$$

$$= \frac{\|\mathbf{x}\| \|\mathbf{y}\| \cos \theta}{\|\mathbf{y}\|}$$

$$= \frac{\mathbf{x} \cdot \mathbf{y}}{\|\mathbf{y}\|}$$

$$= \frac{(2)(2) + (2)(6) + (4)(3)}{\sqrt{2^2 + 6^2 + 3^2}}$$

$$= 4$$

Therefore,

$$\mathbf{proj}_y \mathbf{x} = (\text{comp}_y \, \mathbf{x})\hat{\mathbf{y}} = 4\hat{\mathbf{y}} = 4\frac{\mathbf{y}}{\|\mathbf{y}\|} = 4 \cdot \frac{(2,\ 6,\ 3)}{7} = (\tfrac{8}{7},\ \tfrac{24}{7},\ \tfrac{12}{7})$$

See Figure 29.

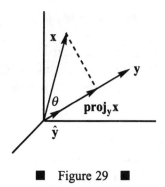

■ Figure 29 ■

Lines. A line is determined by two distinct points, say **p** and **q**. If the vector **pq** is drawn from **p** to **q**, then **pq** defines the line's direction. The description of a line can therefore be re-formulated as follows: A line L is uniquely determined by

• a given point through which L passes,

and

• a given (nonzero) vector which is parallel to L

Let **p** be the given point through which the line will pass, and let **v** be the vector that defines its direction.

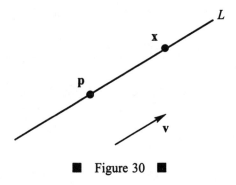

■ Figure 30 ■

From Figure 30, it is easy to see that a point **x** will be on the line if and only if the vector **px** is parallel (or antiparallel) to **v**, which happens if **px** is a scalar multiple of **v**:

$$\mathbf{px} = t\mathbf{v}$$

Or, since **px** = **x** − **p**,

$$\boxed{\mathbf{x} = \mathbf{p} + t\mathbf{v}}$$

This is the parametric equation for the line through **p** parallel to **v**. The scalar t is the **parameter**, and every point on the line is given by a particular choice of t.

Example 25: Find the equation of the line L in $\mathbf{R}^3$ that passes through the point $\mathbf{p} = (2, 4, 2)$ and is parallel to the vector $\mathbf{v} = (1, 2, 3)$. Where does this line pierce the x-y plane?

A point $\mathbf{x} = (x, y, z)$ is on the line L if and only if the vector $\mathbf{px}$ is a scalar multiple of $\mathbf{v}$:

$$\mathbf{px} = t\mathbf{v}$$
$$\mathbf{x} - \mathbf{p} = t\mathbf{v}$$
$$\mathbf{x} = \mathbf{p} + t\mathbf{v}$$
$$\mathbf{x} = (2, 4, 2) + t(1, 2, 3) = (2+t, \ 4+2t, \ 2+3t)$$

Therefore,

$$L = \{(x, y, z): \ x = 2+t, \ y = 4+2t, \ z = 2+3t, \ \text{for any } t \text{ in } \mathbf{R}\}$$

Now, the line intersects the x-y plane when $z = 0$. Since

$$z = 0 \quad \Leftrightarrow \quad 2+3t = 0 \quad \Leftrightarrow \quad t = -\tfrac{2}{3}$$

L pierces the x-y plane when the parameter t takes on the value $-2/3$. For this t,

$$x = 2+t = 2 - \tfrac{2}{3} = \tfrac{4}{3} \quad \text{and} \quad y = 4+2t = 4 - \tfrac{4}{3} = \tfrac{8}{3}$$

so the point of intersection of L and the x-y plane is $\mathbf{a} = (4/3, 8/3, 0)$. See Figure 31.

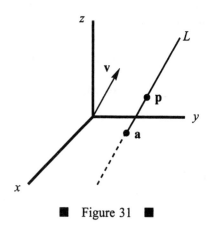

■ Figure 31 ■

Example 26: Give the equation of the line in $\mathbf{R}^4$ that passes through the points $\mathbf{a} = (-1, 1, 2, 0)$ and $\mathbf{b} = (3, 4, 0, -5)$. Does the point $\mathbf{c} = (7, 7, -2, -2)$ lie on this line?

Since the line is parallel to the vector

$$\mathbf{v} = \mathbf{ab} = \mathbf{b} - \mathbf{a} = (3, 4, 0, -5) - (-1, 1, 2, 0) = (4, 3, -2, -5)$$

every point $\mathbf{x}$ on the line is described by the parametric equation

$$\mathbf{x} = \mathbf{a} + t\mathbf{v} = (-1, 1, 2, 0) + t(4, 3, -2, -5)$$
$$= (-1 + 4t, 1 + 3t, 2 - 2t, -5t)$$

The point $\mathbf{c} = (7, 7, -2, -2)$ will lie on this line if and only if there is a value of the parameter t such that

$$(-1 + 4t, 1 + 3t, 2 - 2t, -5t) = (7, 7, -2, -2) \quad (*)$$

However, although the first three components in (*) agree when $t = 2$, the fourth components do not. Therefore, the point $\mathbf{c}$ does not lie on the line. ■

Planes. A plane in $\mathbf{R}^3$ is determined by three noncollinear points. If these points are labeled **a**, **b**, and **c**, then the cross product of the vectors **ab** and **ac** will give a vector **v** perpendicular to the plane. This vector **v** defines the plane's orientation in space; see Figure 32.

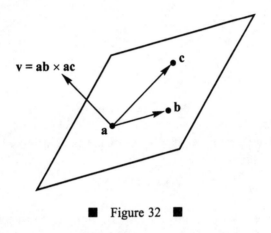

$\mathbf{v} = \mathbf{ab} \times \mathbf{ac}$

■ Figure 32 ■

The description of a plane can be formulated as follows: A plane P is uniquely determined by

- a given point through which P passes,

and

- a given nonzero vector which is **normal**—that is, perpendicular—to P

Let **a** be the given point on the plane, and let **v** be the defining normal vector, with its initial point at **a**. Then, as illustrated in Figure 33, for a point **x** to lie in P, **v** must be perpendicular to the vector **ax**.

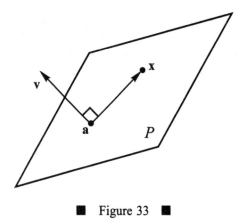

■ Figure 33 ■

Since $\mathbf{v} \perp \mathbf{ax} \Rightarrow \mathbf{v} \cdot \mathbf{ax} = 0$, the plane is determined by the equation

$$\mathbf{v} \cdot \mathbf{ax} = 0 \quad (*)$$

To illustrate, let the point $\mathbf{a} = (a_1, a_2, a_3)$ and the vector $\mathbf{v} = (v_1, v_2, v_3)$. Since for any point $\mathbf{x} = (x, y, z)$, the vector $\mathbf{ax} = \mathbf{x} - \mathbf{a} = (x - a_1, y - a_2, z - a_3)$, equation (*) becomes

$$v_1(x - a_1) + v_2(y - a_2) + v_3(z - a_3) = 0$$

which can also be written as

$$\boxed{v_1 x + v_2 y + v_3 z = d}$$

where $d = v_1 a_1 + v_2 a_2 + v_3 a_3$. For a plane in $\mathbf{R}^3$, this is the **standard equation**. *Note carefully that the coefficients of x, y, and z in the standard equation are precisely the components of a vector normal to the plane.*

Example 27: Give the standard equation of the plane P determined by the points $\mathbf{p} = (2, -1, 2)$, $\mathbf{q} = (2, 2, -1)$, and $\mathbf{r} = (0, 1, 1)$. Does this plane contain the origin? If not, give the equation of the plane which is parallel to P that does contain the origin.

The vectors $\mathbf{pq} = (0, 3, -3)$ and $\mathbf{pr} = (-2, 2, -1)$ lie in P, and their cross product, $\mathbf{pq} \times \mathbf{pr}$, is normal to P. See Figure 34.

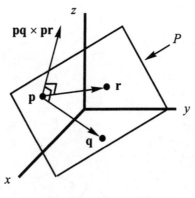

■ Figure 34 ■

Recalling the expression

$$\mathbf{x} \times \mathbf{y} = (x_1, x_2, x_3) \times (y_1, y_2, y_3)$$
$$= (x_2 y_3 - x_3 y_2, \ x_3 y_1 - x_1 y_3, \ x_1 y_2 - x_2 y_1)$$

for the cross product, it is determined that

$$\mathbf{pq} \times \mathbf{pr} = (0, 3, -3) \times (-2, 2, -1)$$
$$= (3 \cdot (-1) - (-3) \cdot 2, \ (-3) \cdot (-2) - 0 \cdot (-1), \ 0 \cdot 2 - 3 \cdot (-2))$$
$$= (3, 6, 6)$$

Since the vector $\mathbf{v} = (3, 6, 6)$ is normal to P, the standard equation of P is given by

$$3x + 6y + 6z = d$$

for some constant d. Substituting the coordinates of any of the three given points ($\mathbf{p}$, $\mathbf{q}$, or $\mathbf{r}$) into this equation yields $d = 12$. Thus, P is given by the equation $3x + 6y + 6z = 12$, or more simply,

$$x + 2y + 2z = 4$$

Now, since $(0, 0, 0)$ does not satisfy this equation, P does not contain the origin. However, the equation $x + 2y + 2z = 0$ specifies a plane parallel to P which *is* satisfied by $\mathbf{0} = (0, 0, 0)$. See Figure 35.

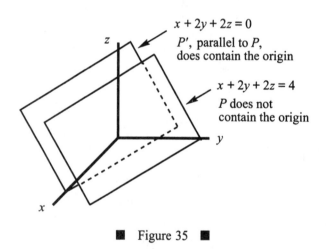

■ Figure 35 ■

$\mathbf{M}$uch of the machinery of linear algebra involves *matrices*, which are rectangular arrays of numbers. In this chapter, the fundamental definitions and operations involving matrices will be stated and illustrated. The rest of the book will then be primarily devoted to applying what is learned here.

Matrices

A rectangular array of numbers, enclosed in a large pair of either parentheses or brackets, such as

$$\begin{pmatrix} 1 & 0 & -3 \\ -2 & 4 & 1 \end{pmatrix} \text{ or } \begin{bmatrix} 1 & 0 & -3 \\ -2 & 4 & 1 \end{bmatrix}$$

is called a **matrix**. The **size** or **dimensions** of a matrix are specified by stating the number of rows and the number of columns it contains. If the matrix consists of m rows and n columns, it is said to be an $\boldsymbol{m}$ **by** $\boldsymbol{n}$ (written $m \times n$) matrix. For example, the matrices above are 2 by 3, since they contain 2 rows and 3 columns:

$$\begin{array}{l} \text{row 1} \longrightarrow \\ \text{row 2} \longrightarrow \end{array} \begin{bmatrix} 1 & 0 & -3 \\ -2 & 4 & 1 \end{bmatrix}$$
$$\uparrow \quad \uparrow \quad \uparrow$$
$$\text{columns 1, 2, 3}$$

Note that the rows are counted from top to bottom, and the columns are counted from left to right.

The numbers in the array are called the **entries** of the matrix, and the location of a particular entry is specified by giving first the row and then the column where it resides. The entry in row i, column j is called the **(i, j) entry**. For example, since the entry -2 in the matrix above is in row 2, column 1, it is the (2, 1) entry. The (1, 2) entry is 0, the (2, 3) entry is 1, and so forth. In general, the (i, j) entry of a matrix A is written a_{ij}, and the statement

$$A = [a_{ij}]_{m \times n}$$

indicates that A is the $m \times n$ matrix whose (i, j) entry is a_{ij}.

Example 1: The set of all $m \times n$ matrices whose entries are real numbers is denoted $M_{m \times n}(\mathbf{R})$. If $A \in M_{2 \times 3}(\mathbf{R})$, how many entries does the matrix A contain?

Since every matrix in $M_{2 \times 3}(\mathbf{R})$ consists of 2 rows and 3 columns, A will contain $2 \times 3 = 6$ entries. An example of such a matrix is

$$A = \begin{bmatrix} 1 & 0 & -3 \\ -2 & 4 & 1 \end{bmatrix} \quad \blacksquare$$

Example 2: If B is the 2×2 matrix whose (i, j) entry is given by the formula $b_{ij} = (-1)^{i+j}(i + j)$, explicitly determine B.

The (1, 1) entry of B is $b_{11} = (-1)^{1+1}(1 + 1) = 2$; the (1, 2) entry is $b_{12} = (-1)^{1+2}(1 + 2) = -3$; the (2, 1) entry is also -3; and the (2, 2) entry is $b_{22} = (-1)^{2+2}(2 + 2) = 4$. Therefore,

$$B = \begin{bmatrix} b_{11} & b_{12} \\ b_{21} & b_{22} \end{bmatrix} = \begin{bmatrix} 2 & -3 \\ -3 & 4 \end{bmatrix}$$ ∎

Example 3: Give the 3×3 matrix whose (i, j) entry is expressed by the formula

$$\delta_{ij} = \begin{cases} 1 & \text{if } i = j \\ 0 & \text{if } i \neq j \end{cases}$$

(δ_{ij} is called the *Kronecker delta*.)

The $(1, 1)$, $(2, 2)$, and $(3, 3)$ entries are each equal to 1, but all other entries are 0. Thus, the matrix is

$$\left[\delta_{ij} \right]_{3 \times 3} = \begin{bmatrix} 1 & 0 & 0 \\ 0 & 1 & 0 \\ 0 & 0 & 1 \end{bmatrix}$$ ∎

Entries along the diagonal. Any entry whose column number matches its row number is called a **diagonal entry**; all other entries are called **off-diagonal**. The diagonal entries in each of the following matrices are highlighted:

$$A = \begin{bmatrix} \boxed{1} & 0 & -3 \\ -2 & \boxed{4} & 1 \end{bmatrix}, \quad B = \begin{bmatrix} \boxed{2} & -3 \\ -3 & \boxed{4} \end{bmatrix}, \quad \left[\delta_{ij} \right]_{3 \times 3} = \begin{bmatrix} \boxed{1} & 0 & 0 \\ 0 & \boxed{1} & 0 \\ 0 & 0 & \boxed{1} \end{bmatrix}$$

In the matrix A, the diagonal entries are $a_{11} = 1$ and $a_{22} = 4$; in B, the diagonal entries are $b_{11} = 2$ and $b_{22} = 4$; and in the matrix $[\delta_{ij}]_{3 \times 3}$, the diagonal entries are $\delta_{11} = \delta_{22} = \delta_{33} = 1$.

If every off-diagonal entry of a matrix equals zero, then the matrix is called a **diagonal matrix**. For example, the matrix $[\delta_{ij}]_{3\times3}$ above is a diagonal matrix. It is not uncommon in such cases, particularly with large matrices, to simply leave *blank* any entry that equals zero. For example,

$$\begin{bmatrix} 1 & 0 & 0 \\ 0 & 1 & 0 \\ 0 & 0 & 1 \end{bmatrix} \quad \text{and} \quad \begin{bmatrix} 1 & & \\ & 1 & \\ & & 1 \end{bmatrix}$$

are two ways of writing exactly the same matrix. Blocks of zeros are often left blank in nondiagonal matrices also. An $n \times n$ diagonal matrix whose entries—from the upper-left to the lower-right—are $a_{11}, a_{22}, \ldots, a_{nn}$ is often written Diag(a_{11}, $a_{22}, \ldots, a_{nn}$).

Square matrices. Any matrix which has as many columns as rows is called a **square matrix**. The 2×2 matrix in Example 2 and the 3×3 matrix in Example 3 are square. If a square matrix has n rows and n columns, that is, if its size is $n \times n$, then the matrix is said to be **of order n**.

Triangular matrices. If all the entries below the diagonal of a square matrix are zero, then the matrix is said to be **upper triangular**. The following matrix, U, is an example of an upper triangular matrix of order 3:

$$U = \begin{bmatrix} 1 & -3 & 2 \\ 0 & 4 & 1 \\ 0 & 0 & -1 \end{bmatrix} = \begin{bmatrix} 1 & -3 & 2 \\ & 4 & 1 \\ & & -1 \end{bmatrix}$$

If all the entries above the diagonal of a square matrix are zero, then the matrix is said to be **lower triangular**. The following matrix, L, is an example of a lower triangular matrix of order 4:

$$L = \begin{bmatrix} 2 & 0 & 0 & 0 \\ 1 & 4 & 0 & 0 \\ 0 & 5 & -1 & 0 \\ 1 & -2 & 1 & -3 \end{bmatrix} = \begin{bmatrix} 2 & & & \\ 1 & 4 & & \\ 0 & 5 & -1 & \\ 1 & -2 & 1 & -3 \end{bmatrix}$$

A matrix is called **triangular** if it is either upper triangular or lower triangular. A diagonal matrix is one that is both upper and lower triangular.

The transpose of a matrix. One of the most basic operations that can be performed on a matrix is to form its transpose. Let A be a matrix; then the **transpose** of A, a matrix ·denoted by A^T, is obtained by writing the rows of A as columns. More precisely, row i of A is column i of A^T (which implies that column j of A is row j of A^T). If A is $m \times n$, then A^T will be $n \times m$. Also, it follows immediately from the definition that $(A^T)^T = A$.

Example 4: The transpose of the 2×3 matrix

$$A = \begin{bmatrix} 1 & 0 & -3 \\ -2 & 4 & 1 \end{bmatrix}$$

is the 3×2 matrix

$$A^T = \begin{bmatrix} 1 & -2 \\ 0 & 4 \\ -3 & 1 \end{bmatrix} \quad \blacksquare$$

Example 5: Note that each of the matrices in Examples 2 and 3 is equal to its own transpose:

$$\begin{bmatrix} 2 & -3 \\ -3 & 4 \end{bmatrix}^{\mathrm{T}} = \begin{bmatrix} 2 & -3 \\ -3 & 4 \end{bmatrix} \quad \text{and} \quad \begin{bmatrix} 1 & 0 & 0 \\ 0 & 1 & 0 \\ 0 & 0 & 1 \end{bmatrix}^{\mathrm{T}} = \begin{bmatrix} 1 & 0 & 0 \\ 0 & 1 & 0 \\ 0 & 0 & 1 \end{bmatrix}$$

Any matrix which equals its own transpose is called a **symmetric** matrix. ■

Row and column matrices. A matrix that consists of precisely one row is called a **row matrix**, and a matrix that consists of precisely one column is called a **column matrix**. For example,

$$R = \begin{bmatrix} 1 & 0 & -3 \end{bmatrix}$$

is a row matrix, while

$$C = \begin{bmatrix} 1 \\ -2 \\ 0 \\ 4 \end{bmatrix}$$

is a column matrix. Note that the transpose of a row matrix is a column matrix and vice versa; for example,

$$R^{\mathrm{T}} = \begin{bmatrix} 1 \\ 0 \\ -3 \end{bmatrix} \quad \text{and} \quad C^{\mathrm{T}} = \begin{bmatrix} 1 & -2 & 0 & 4 \end{bmatrix}$$

Row and column matrices provide alternate notations for a vector. For example, the vector $\mathbf{v} = (2, -1, 6)$ in $\mathbf{R}^3$ can be expressed as either a row matrix or a column matrix:

$$\mathbf{v} = \begin{bmatrix} 2 & -1 & 6 \end{bmatrix} \quad \text{or} \quad \mathbf{v} = \begin{bmatrix} 2 \\ -1 \\ 6 \end{bmatrix}$$

It is common to denote such a matrix by a bold, lower-case (rather than an italic, upper-case) letter and to refer to it as either a **row vector** or a **column vector**.

Zero matrices. Any matrix all of whose entries are zero is called a **zero matrix** and is generically denoted *0*. If it is important to explicitly indicate the size of a zero matrix, then subscript notation is used. For example, the 2×3 zero matrix

$$\begin{bmatrix} 0 & 0 & 0 \\ 0 & 0 & 0 \end{bmatrix}$$

would be written $0_{2 \times 3}$. If a zero matrix is a row or column matrix, it is usually denoted **0**, which is consistent with the designation of **0** as the zero vector.

Operations with Matrices

As far as linear algebra is concerned, the two most important operations with vectors are vector addition [adding two (or more) vectors] and scalar multiplication (multiplying a vector by a scalar). Analogous operations are defined for matrices.

Matrix addition. If *A* and *B* are matrices *of the same size*, then they can be added. (This is similar to the restriction on adding vectors, namely, only vectors from the same space $\mathbf{R}^n$

can be added; you cannot add a 2-vector to a 3-vector, for example.) If $A = [a_{ij}]$ and $B = [b_{ij}]$ are both $m \times n$ matrices, then their sum, $C = A + B$, is also an $m \times n$ matrix, and its entries are given by the formula

$$c_{ij} = a_{ij} + b_{ij}$$

Thus, to find the entries of $A + B$, simply add the corresponding entries of A and B.

Example 6: Consider the following matrices:

$$F = \begin{bmatrix} 2 & -1 \\ 3 & 0 \\ -5 & 2 \end{bmatrix}, \quad G = \begin{bmatrix} 4 & 4 & -3 \\ 0 & -1 & -2 \end{bmatrix}, \quad H = \begin{bmatrix} 1 & 6 \\ -1 & -2 \\ 0 & -3 \end{bmatrix}$$

Which two can be added? What is their sum?

Since only matrices of the same size can be added, only the sum $F + H$ is defined (G cannot be added to either F or H). The sum of F and H is

$$F + H = \begin{bmatrix} 2 & -1 \\ 3 & 0 \\ -5 & 2 \end{bmatrix} + \begin{bmatrix} 1 & 6 \\ -1 & -2 \\ 0 & -3 \end{bmatrix} = \begin{bmatrix} 2+1 & -1+6 \\ 3-1 & 0-2 \\ -5+0 & 2-3 \end{bmatrix} = \begin{bmatrix} 3 & 5 \\ 2 & -2 \\ -5 & -1 \end{bmatrix} \quad \blacksquare$$

Since addition of real numbers is commutative, it follows that addition of matrices (when it is defined) is also commutative; that is, for any matrices A and B of the same size, $A + B$ will always equal $B + A$.

Example 7: If any matrix A is added to the zero matrix of the same size, the result is clearly equal to A:

$$A + 0 = 0 + A = A$$

This is the matrix analog of the statement $a + 0 = 0 + a = a$, which expresses the fact that the number 0 is the additive identity in the set of real numbers. ■

Example 8: Find the matrix B such that $A + B = C$, where

$$A = \begin{bmatrix} 2 & 0 \\ 1 & 4 \end{bmatrix} \quad \text{and} \quad C = \begin{bmatrix} 3 & -1 \\ -2 & 2 \end{bmatrix}$$

If

$$B = \begin{bmatrix} b_{11} & b_{12} \\ b_{21} & b_{22} \end{bmatrix}$$

then the matrix equation $A + B = C$ becomes

$$\begin{bmatrix} 2 + b_{11} & 0 + b_{12} \\ 1 + b_{21} & 4 + b_{22} \end{bmatrix} = \begin{bmatrix} 3 & -1 \\ -2 & 2 \end{bmatrix}$$

Since two matrices are equal if and only if they are of the same size and their corresponding entries are equal, this last equation implies

$$b_{11} = 1, \quad b_{12} = -1, \quad b_{21} = -3, \quad \text{and} \quad b_{22} = -2$$

Therefore,

$$B = \begin{bmatrix} 1 & -1 \\ -3 & -2 \end{bmatrix}$$

This example motivates the definition of matrix *subtraction*: If A and B are matrices of the same size, then the entries of $A - B$ are found by simply subtracting the entries of B from the corresponding entries of A. Since the equation $A + B = C$ is equivalent to $B = C - A$, employing matrix subtraction above would yield the same result:

$$B = C - A = \begin{bmatrix} 3 & -1 \\ -2 & 2 \end{bmatrix} - \begin{bmatrix} 2 & 0 \\ 1 & 4 \end{bmatrix} = \begin{bmatrix} 3-2 & -1-0 \\ -2-1 & 2-4 \end{bmatrix} = \begin{bmatrix} 1 & -1 \\ -3 & -2 \end{bmatrix} \quad \blacksquare$$

Scalar multiplication. A matrix can be multiplied by a scalar as follows. If $A = [a_{ij}]$ is a matrix and k is a scalar, then

$$kA = [ka_{ij}]$$

That is, the matrix kA is obtained by multiplying each entry of A by k.

Example 9: If

$$A = \begin{bmatrix} 1 & 0 & -3 \\ -2 & 4 & 1 \end{bmatrix}$$

then the scalar multiple $2A$ is obtained by multiplying every entry of A by 2:

$$2A = \begin{bmatrix} 2 & 0 & -6 \\ -4 & 8 & 2 \end{bmatrix} \quad \blacksquare$$

Example 10: If A and B are matrices of the same size, then $A - B = A + (-B)$, where $-B$ is the scalar multiple $(-1)B$. If

$$A = \begin{bmatrix} 1 & 0 & -3 \\ -2 & 4 & 1 \end{bmatrix} \quad \text{and} \quad B = \begin{bmatrix} 3 & -2 & 0 \\ 1 & -1 & -5 \end{bmatrix}$$

then

$$A - B = A + (-B) = \begin{bmatrix} 1 & 0 & -3 \\ -2 & 4 & 1 \end{bmatrix} + \begin{bmatrix} -3 & 2 & 0 \\ -1 & 1 & 5 \end{bmatrix} = \begin{bmatrix} -2 & 2 & -3 \\ -3 & 5 & 6 \end{bmatrix}$$

This definition of matrix subtraction is consistent with the definition illustrated in Example 8. ∎

Example 11: If

$$A = \begin{bmatrix} 1 & 0 & -3 \\ -2 & 4 & 1 \end{bmatrix} \quad \text{and} \quad B = \begin{bmatrix} 1 & \frac{1}{2} \\ 2 & -4 \\ 3 & -1 \end{bmatrix}$$

then

$$3A^{\mathrm{T}} + 4B = \begin{bmatrix} 3 & -6 \\ 0 & 12 \\ -9 & 3 \end{bmatrix} + \begin{bmatrix} 4 & 2 \\ 8 & -16 \\ 12 & -4 \end{bmatrix} = \begin{bmatrix} 7 & -4 \\ 8 & -4 \\ 3 & -1 \end{bmatrix} \quad \blacksquare$$

Matrix multiplication. By far the most important operation involving matrices is *matrix multiplication*, the process of multiplying one matrix by another. The first step in defining matrix multiplication is to recall the definition of the dot product of two vectors. Let **r** and **c** be two n-vectors. Writing **r** as a $1 \times n$ row matrix and **c** as an $n \times 1$ column matrix, the dot product of **r** and **c** is

$$\mathbf{r} \cdot \mathbf{c} = \begin{bmatrix} r_1 & r_2 & \cdots & r_n \end{bmatrix} \cdot \begin{bmatrix} c_1 \\ c_2 \\ \vdots \\ c_n \end{bmatrix} = r_1 c_1 + r_2 c_2 + \cdots + r_n c_n$$

Note that in order for the dot product of **r** and **c** to be defined, both must contain the same number of entries. Also, the order in which these matrices are written in this product is important here: The row vector comes first, the column vector second.

Now, for the final step: How are two general matrices multiplied? First, in order to form the product AB, *the number of columns of A must match the number of rows of B*; if this condition does not hold, then the product AB is not defined. This criterion follows from the restriction stated above for multiplying a row matrix **r** by a column matrix **c**, namely that the number of entries in **r** must match the number of entries in **c**. If A is $m \times n$ and B is $n \times p$, then the product AB is defined, and the size of the product matrix AB will be $m \times p$. The following diagram is helpful in determining if a matrix product is defined, and if so, the dimensions of the product:

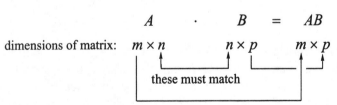

Thinking of the $m \times n$ matrix A as composed of the row vectors $\mathbf{r}_1, \mathbf{r}_2, \ldots, \mathbf{r}_m$ from $\mathbf{R}^n$ and the $n \times p$ matrix B as composed of the column vectors $\mathbf{c}_1, \mathbf{c}_2, \ldots, \mathbf{c}_p$ from $\mathbf{R}^n$,

$$A = \begin{bmatrix} — & \text{row } 1 & \rightarrow \\ — & \text{row } 2 & \rightarrow \\ & \cdots & \\ — & \text{row } m & \rightarrow \end{bmatrix} = \begin{bmatrix} — & \mathbf{r}_1 & \rightarrow \\ — & \mathbf{r}_2 & \rightarrow \\ & \cdots & \\ — & \mathbf{r}_m & \rightarrow \end{bmatrix}$$

and

$$B = \begin{bmatrix} | & | & & | \\ \text{col } 1 & \text{col } 2 & \vdots & \text{col } p \\ \downarrow & \downarrow & & \downarrow \end{bmatrix} = \begin{bmatrix} | & | & & | \\ \mathbf{c}_1 & \mathbf{c}_2 & \vdots & \mathbf{c}_p \\ \downarrow & \downarrow & & \downarrow \end{bmatrix}$$

the rule for computing the entries of the matrix product AB is $\mathbf{r}_i \cdot \mathbf{c}_j = (AB)_{ij}$, that is,

> The dot product of row i in A and column
> j in B gives the (i, j) entry of AB

Example 12: Given the two matrices

$$A = \begin{bmatrix} 1 & 0 & -3 \\ -2 & 4 & 1 \end{bmatrix} \quad \text{and} \quad B = \begin{bmatrix} 1 & 0 & 4 & 1 \\ -2 & 3 & -1 & 5 \\ 0 & -1 & 2 & 1 \end{bmatrix}$$

determine which matrix product, AB or BA, is defined and evaluate it.

Since A is 2×3 and B is 3×4, the product AB, in that order, is defined, and the size of the product matrix AB will be 2×4. The product BA is *not* defined, since the first factor (B) has 4 columns but the second factor (A) has only 2 rows. The number of columns of the first matrix must match the number of rows of the second matrix in order for their product to be defined.

Taking the dot product of row 1 in A and column 1 in B gives the (1, 1) entry in AB. Since

$$\begin{bmatrix} 1 & 0 & -3 \end{bmatrix} \cdot \begin{bmatrix} 1 \\ -2 \\ 0 \end{bmatrix} = (1)(1) + (0)(-2) + (-3)(0) = 1$$

the (1, 1) entry in AB is 1:

$$\begin{bmatrix} \boxed{\begin{matrix} 1 & 0 & -3 \end{matrix}} \\ -2 & 4 & 1 \end{bmatrix} \begin{bmatrix} \boxed{\begin{matrix} 1 \\ -2 \\ 0 \end{matrix}} & \begin{matrix} 0 & 4 & 1 \\ 3 & -1 & 5 \\ -1 & 2 & 1 \end{matrix} \end{bmatrix} = \begin{bmatrix} \boxed{1} \end{bmatrix}$$

The dot product of row 1 in A and column 2 in B gives the (1, 2) entry in AB,

$$\begin{bmatrix} \boxed{\begin{matrix} 1 & 0 & -3 \end{matrix}} \\ -2 & 4 & 1 \end{bmatrix} \begin{bmatrix} 1 & \boxed{\begin{matrix} 0 \\ 3 \\ -1 \end{matrix}} & \begin{matrix} 4 & 1 \\ -1 & 5 \\ 2 & 1 \end{matrix} \end{bmatrix} = \begin{bmatrix} 1 & \boxed{3} \end{bmatrix}$$

and the dot product of row 1 in A and column 3 in B gives the (1, 3) entry in AB:

$$\begin{bmatrix} \boxed{\begin{matrix} 1 & 0 & -3 \end{matrix}} \\ -2 & 4 & 1 \end{bmatrix} \begin{bmatrix} 1 & 0 & \boxed{\begin{matrix} 4 \\ -1 \\ 2 \end{matrix}} & 1 \\ -2 & 3 & & 5 \\ 0 & -1 & & 1 \end{bmatrix} = \begin{bmatrix} 1 & 3 & \boxed{-2} \end{bmatrix}$$

The first row of the product is completed by taking the dot product of row 1 in A and column 4 in B, which gives the (1, 4) entry in AB:

$$\left[\begin{array}{|ccc|} \hline 1 & 0 & -3 \\ \hline -2 & 4 & 1 \end{array}\right] \left[\begin{array}{ccc|c} 1 & 0 & 4 & \boxed{1} \\ -2 & 3 & -1 & 5 \\ 0 & -1 & 2 & 1 \end{array}\right] = \left[\begin{array}{cccc} 1 & 3 & -2 & \boxed{-2} \end{array}\right]$$

Now for the second row of AB: The dot product of row 2 in A and column 1 in B gives the $(2, 1)$ entry in AB,

$$\left[\begin{array}{ccc} 1 & 0 & -3 \\ \hline -2 & 4 & 1 \end{array}\right] \left[\begin{array}{c|ccc} \boxed{\begin{array}{c}1\\-2\\0\end{array}} & \begin{array}{c}0\\3\\-1\end{array} & \begin{array}{c}4\\-1\\2\end{array} & \begin{array}{c}1\\5\\1\end{array} \end{array}\right] = \left[\begin{array}{cccc} 1 & 3 & -2 & -2 \\ \boxed{-10} & & & \end{array}\right]$$

and the dot product of row 2 in A and column 2 in B gives the $(2, 2)$ entry in AB:

$$\left[\begin{array}{ccc} 1 & 0 & -3 \\ \hline -2 & 4 & 1 \end{array}\right] \left[\begin{array}{c|c|cc} 1 & \boxed{0} & 4 & 1 \\ -2 & \boxed{3} & -1 & 5 \\ 0 & \boxed{-1} & 2 & 1 \end{array}\right] = \left[\begin{array}{cccc} 1 & 3 & -2 & -2 \\ -10 & \boxed{11} & & \end{array}\right]$$

Finally, taking the dot product of row 2 in A with columns 3 and 4 in B gives (respectively) the $(2, 3)$ and $(2, 4)$ entries in AB:

$$\left[\begin{array}{ccc} 1 & 0 & -3 \\ \hline -2 & 4 & 1 \end{array}\right] \left[\begin{array}{cc|c|c} 1 & 0 & \boxed{4} & 1 \\ -2 & 3 & \boxed{-1} & 5 \\ 0 & -1 & \boxed{2} & 1 \end{array}\right] = \left[\begin{array}{cccc} 1 & 3 & -2 & -2 \\ -10 & 11 & \boxed{-10} & \end{array}\right]$$

$$\left[\begin{array}{ccc} 1 & 0 & -3 \\ \hline -2 & 4 & 1 \end{array}\right] \left[\begin{array}{ccc|c} 1 & 0 & 4 & \boxed{1} \\ -2 & 3 & -1 & \boxed{5} \\ 0 & -1 & 2 & \boxed{1} \end{array}\right] = \left[\begin{array}{cccc} 1 & 3 & -2 & -2 \\ -10 & 11 & -10 & \boxed{19} \end{array}\right]$$

Therefore,

$$AB = \begin{bmatrix} 1 & 0 & -3 \\ -2 & 4 & 1 \end{bmatrix} \begin{bmatrix} 1 & 0 & 4 & 1 \\ -2 & 3 & -1 & 5 \\ 0 & -1 & 2 & 1 \end{bmatrix} = \begin{bmatrix} 1 & 3 & -2 & -2 \\ -10 & 11 & -10 & 19 \end{bmatrix}$$ ∎

Example 13: If

$$C = \begin{bmatrix} 0 & -2 & 1 & 5 & 3 \\ 1 & 6 & -4 & 8 & 2 \\ -3 & 2 & 0 & -1 & -1 \\ 0 & 10 & 7 & 1 & 3 \end{bmatrix}$$

and

$$D = \begin{bmatrix} 4 & 12 & 0 & -2 & -7 & -3 \\ 0 & -1 & 3 & 5 & -2 & 1 \\ 2 & 2 & -4 & 0 & 9 & 4 \\ 6 & -5 & 2 & 3 & -1 & 0 \\ -3 & 4 & 1 & -1 & 8 & -1 \end{bmatrix}$$

compute the (3, 5) entry of the product CD.

First, note that since C is 4×5 and D is 5×6, the product CD is indeed defined, and its size is 4×6. However, there is no need to compute all twenty-four entries of CD if only one particular entry is desired. The (3, 5) entry of CD is the dot product of row 3 in C and column 5 in D:

$$\begin{bmatrix} -3 & 2 & 0 & -1 & -1 \end{bmatrix} \cdot \begin{bmatrix} -7 \\ -2 \\ 9 \\ -1 \\ 8 \end{bmatrix} = \begin{aligned} &(-3)(-7)+(2)(-2)+(0)(9) \\ &\quad +(-1)(-1)+(-1)(8)=10 \end{aligned}$$

■

Example 14: If

$$A = \begin{bmatrix} 1 & 0 & -3 \\ -2 & 4 & 1 \end{bmatrix} \quad \text{and} \quad B = \begin{bmatrix} 2 & -1 \\ 3 & 0 \\ -5 & 2 \end{bmatrix}$$

verify that

$$AB = \begin{bmatrix} 1 & 0 & -3 \\ -2 & 4 & 1 \end{bmatrix} \begin{bmatrix} 2 & -1 \\ 3 & 0 \\ -5 & 2 \end{bmatrix} = \begin{bmatrix} 17 & -7 \\ 3 & 4 \end{bmatrix}$$

but

$$BA = \begin{bmatrix} 2 & -1 \\ 3 & 0 \\ -5 & 2 \end{bmatrix} \begin{bmatrix} 1 & 0 & -3 \\ -2 & 4 & 1 \end{bmatrix} = \begin{bmatrix} 4 & -4 & -7 \\ 3 & 0 & -9 \\ -9 & 8 & 17 \end{bmatrix}$$

In particular, note that even though both products AB and BA are defined, AB does not equal BA; indeed, they're not even the same size!   ■

The previous example gives one illustration of what is perhaps the most important distinction between the multiplication of scalars and the multiplication of matrices. For real numbers a and b, the equation $ab = ba$ always holds, that is, multiplication of real numbers is commutative; the order in which the factors are written is irrelevant. However, it is decidedly false that matrix multiplication is commutative. For the matrices A and B given in Example 14, both products AB and BA were defined, but they certainly were not identical. In fact, the matrix AB was 2×2, while the matrix BA was 3×3. Here is another illustration of the noncommutativity of matrix multiplication: Consider the matrices

$$C = \begin{bmatrix} 1 & -2 \\ 0 & 4 \\ -3 & 1 \end{bmatrix} \quad \text{and} \quad D = \begin{bmatrix} 2 & -1 \\ 3 & 0 \end{bmatrix}$$

Since C is 3×2 and D is 2×2, the product CD is defined, its size is 3×2, and

$$CD = \begin{bmatrix} 1 & -2 \\ 0 & 4 \\ -3 & 1 \end{bmatrix} \begin{bmatrix} 2 & -1 \\ 3 & 0 \end{bmatrix} = \begin{bmatrix} -4 & -1 \\ 12 & 0 \\ -3 & 3 \end{bmatrix}$$

The product DC, however, is not defined, since the number of columns of D (which is 2) does not equal the number of rows of C (which is 3). Therefore, $CD \neq DC$, since DC doesn't even exist.

Because of the sensitivity to the order in which the factors are written, one does not typically say simply, "Multiply the matrices A and B." It is usually important to indicate which matrix comes first and which comes second in the product. For this reason, the statement "Multiply A on the right by B" means to form the product AB, while "Multiply A on the left by B" means to form the product BA.

Example 15: If

$$A = \begin{bmatrix} 1 & -2 \\ -4 & 5 \end{bmatrix}$$

and **x** is the vector (−2, 3), show how A can be multiplied on the right by **x** and compute the product.

Since A is 2 × 2, in order to multiply A on the right by a matrix, that matrix must have 2 rows. Therefore, if **x** is written as the 2 × 1 *column* matrix

$$\mathbf{x} = \begin{bmatrix} -2 \\ 3 \end{bmatrix}$$

then the product $A\mathbf{x}$ can be computed, and the result is another 2 × 1 column matrix:

$$A\mathbf{x} = \begin{bmatrix} 1 & -2 \\ -4 & 5 \end{bmatrix}\begin{bmatrix} -2 \\ 3 \end{bmatrix} = \begin{bmatrix} -8 \\ 23 \end{bmatrix} \qquad \blacksquare$$

Example 16: Consider the matrices

$$A = \begin{bmatrix} 1 & -2 \\ 3 & 7 \end{bmatrix} \quad \text{and} \quad B = \begin{bmatrix} -1 & 0 \\ 4 & 2 \end{bmatrix}$$

If A is multiplied on the right by B, the result is

$$AB = \begin{bmatrix} 1 & -2 \\ 3 & 7 \end{bmatrix}\begin{bmatrix} -1 & 0 \\ 4 & 2 \end{bmatrix} = \begin{bmatrix} -9 & -4 \\ 25 & 14 \end{bmatrix}$$

but if A is multiplied on the left by B, the result is

$$BA = \begin{bmatrix} -1 & 0 \\ 4 & 2 \end{bmatrix}\begin{bmatrix} 1 & -2 \\ 3 & 7 \end{bmatrix} = \begin{bmatrix} -1 & 2 \\ 10 & 6 \end{bmatrix}$$

Note that both products are defined and of the same size, but they are not equal. ■

Example 17: If A and B are square matrices such that $AB = BA$, then A and B are said to *commute*. Show that any two square diagonal matrices of order 2 commute.

Let

$$A = \begin{bmatrix} a_{11} & 0 \\ 0 & a_{22} \end{bmatrix} \quad \text{and} \quad B = \begin{bmatrix} b_{11} & 0 \\ 0 & b_{22} \end{bmatrix}$$

be two arbitrary 2×2 diagonal matrices. Then

$$AB = \begin{bmatrix} a_{11} & 0 \\ 0 & a_{22} \end{bmatrix}\begin{bmatrix} b_{11} & 0 \\ 0 & b_{22} \end{bmatrix} = \begin{bmatrix} a_{11}b_{11} & 0 \\ 0 & a_{22}b_{22} \end{bmatrix}$$

and

$$BA = \begin{bmatrix} b_{11} & 0 \\ 0 & b_{22} \end{bmatrix}\begin{bmatrix} a_{11} & 0 \\ 0 & a_{22} \end{bmatrix} = \begin{bmatrix} b_{11}a_{11} & 0 \\ 0 & b_{22}a_{22} \end{bmatrix}$$

Since $a_{11}b_{11} = b_{11}a_{11}$ and $a_{22}b_{22} = b_{22}a_{22}$, AB does indeed equal BA, as desired. ■

Although matrix multiplication is usually not commutative, it is *sometimes* commutative; for example, if

$$E = \begin{bmatrix} 1 & 0 \\ 0 & 4 \end{bmatrix} \quad \text{and} \quad F = \begin{bmatrix} -2 & 0 \\ 0 & 3 \end{bmatrix}$$

then

$$EF = \begin{bmatrix} 1 & 0 \\ 0 & 4 \end{bmatrix}\begin{bmatrix} -2 & 0 \\ 0 & 3 \end{bmatrix} = \begin{bmatrix} -2 & 0 \\ 0 & 12 \end{bmatrix} = \begin{bmatrix} -2 & 0 \\ 0 & 3 \end{bmatrix}\begin{bmatrix} 1 & 0 \\ 0 & 4 \end{bmatrix} = FE$$

Despite examples such as these, it must be stated that *in general, matrix multiplication is not commutative.*

There is another difference between the multiplication of scalars and the multiplication of matrices. If a and b are real numbers, then the equation $ab = 0$ implies that $a = 0$ or $b = 0$. That is, the only way a product of real numbers can equal 0 is if at least one of the factors is itself 0. The analogous statement for matrices, however, is not true. For instance, if

$$G = \begin{bmatrix} 1 & 1 \\ 1 & 1 \end{bmatrix} \quad \text{and} \quad H = \begin{bmatrix} 1 & -1 \\ -1 & 1 \end{bmatrix}$$

then

$$GH = \begin{bmatrix} 1 & 1 \\ 1 & 1 \end{bmatrix}\begin{bmatrix} 1 & -1 \\ -1 & 1 \end{bmatrix} = \begin{bmatrix} 0 & 0 \\ 0 & 0 \end{bmatrix} = 0$$

Note that even though neither G nor H is a zero matrix, the product GH is.

Yet another difference between the multiplication of scalars and the multiplication of matrices is the lack of a general cancellation law for matrix multiplication. If a, b, and c are real numbers with $a \neq 0$, then, by canceling out the factor a, the equation $ab = ac$ implies $b = c$. No such law exists for matrix multiplication; that is, the statement $AB = AC$ does *not* imply $B = C$, even if A is nonzero. For example, if

$$A = \begin{bmatrix} 1 & 1 \\ 1 & 1 \end{bmatrix}, \quad B = \begin{bmatrix} 1 & 2 \\ 3 & 4 \end{bmatrix}, \quad \text{and} \quad C = \begin{bmatrix} 3 & 4 \\ 1 & 2 \end{bmatrix}$$

then both

$$AB = \begin{bmatrix} 1 & 1 \\ 1 & 1 \end{bmatrix}\begin{bmatrix} 1 & 2 \\ 3 & 4 \end{bmatrix} = \begin{bmatrix} 4 & 6 \\ 4 & 6 \end{bmatrix}$$

and

$$AC = \begin{bmatrix} 1 & 1 \\ 1 & 1 \end{bmatrix}\begin{bmatrix} 3 & 4 \\ 1 & 2 \end{bmatrix} = \begin{bmatrix} 4 & 6 \\ 4 & 6 \end{bmatrix}$$

Thus, even though $AB = AC$ and A is not a zero matrix, B does not equal C. ∎

Example 18: Although matrix multiplication is not always commutative, it *is* always *associative*. That is, if A, B, and C are any three matrices such that the product $(AB)C$ is defined, then the product $A(BC)$ is also defined, and

$$\boxed{(AB)C = A(BC)}$$

That is, as long as the order of the factors is unchanged, how they are *grouped* is irrelevant.

Verify the associative law for the matrices

$$A = \begin{bmatrix} 1 & 2 \\ 0 & -1 \end{bmatrix}, \quad B = \begin{bmatrix} -1 & 3 & 0 \\ 4 & 1 & -6 \end{bmatrix}, \quad \text{and} \quad C = \begin{bmatrix} 2 \\ 1 \\ -3 \end{bmatrix}$$

First, since

$$AB = \begin{bmatrix} 1 & 2 \\ 0 & -1 \end{bmatrix}\begin{bmatrix} -1 & 3 & 0 \\ 4 & 1 & -6 \end{bmatrix} = \begin{bmatrix} 7 & 5 & -12 \\ -4 & -1 & 6 \end{bmatrix}$$

the product $(AB)C$ is

$$(AB)C = \begin{bmatrix} 7 & 5 & -12 \\ -4 & -1 & 6 \end{bmatrix} \begin{bmatrix} 2 \\ 1 \\ -3 \end{bmatrix} = \begin{bmatrix} 55 \\ -27 \end{bmatrix}$$

Now, since

$$BC = \begin{bmatrix} -1 & 3 & 0 \\ 4 & 1 & -6 \end{bmatrix} \begin{bmatrix} 2 \\ 1 \\ -3 \end{bmatrix} = \begin{bmatrix} 1 \\ 27 \end{bmatrix}$$

the product $A(BC)$ is

$$A(BC) = \begin{bmatrix} 1 & 2 \\ 0 & -1 \end{bmatrix} \begin{bmatrix} 1 \\ 27 \end{bmatrix} = \begin{bmatrix} 55 \\ -27 \end{bmatrix}$$

Therefore, $(AB)C = A(BC)$, as expected. Note that the associative law implies that the product of A, B, and C (in that order) can be written simply as ABC; parentheses are not needed to resolve any ambiguity, because there is no ambiguity. ∎

Example 19: For the matrices

$$A = \begin{bmatrix} 1 & 2 \\ 0 & -1 \end{bmatrix} \quad \text{and} \quad B = \begin{bmatrix} -1 & 3 & 0 \\ 4 & 1 & -6 \end{bmatrix}$$

verify the equation $(AB)^{\text{T}} = B^{\text{T}} A^{\text{T}}$.

First,

$$AB = \begin{bmatrix} 1 & 2 \\ 0 & -1 \end{bmatrix} \begin{bmatrix} -1 & 3 & 0 \\ 4 & 1 & -6 \end{bmatrix} = \begin{bmatrix} 7 & 5 & -12 \\ -4 & -1 & 6 \end{bmatrix}$$

implies

$$(AB)^{\mathrm{T}} = \begin{bmatrix} 7 & -4 \\ 5 & -1 \\ -12 & 6 \end{bmatrix}$$

Now, since

$$B^{\mathrm{T}} A^{\mathrm{T}} = \begin{bmatrix} -1 & 4 \\ 3 & 1 \\ 0 & -6 \end{bmatrix} \begin{bmatrix} 1 & 0 \\ 2 & -1 \end{bmatrix} = \begin{bmatrix} 7 & -4 \\ 5 & -1 \\ -12 & 6 \end{bmatrix}$$

$B^{\mathrm{T}} A^{\mathrm{T}}$ does indeed equal $(AB)^{\mathrm{T}}$. In fact, the equation

$$\boxed{(AB)^{\mathrm{T}} = B^{\mathrm{T}} A^{\mathrm{T}}}$$

holds true for *any* two matrices for which the product AB is defined. This says that if the product AB is defined, then the transpose of the product is equal to the product of the transposes *in the reverse order*. ■

Identity matrices. The zero matrix $0_{m \times n}$ plays the role of the additive identity in the set of $m \times n$ matrices in the same way that the number 0 does in the set of real numbers (recall Example 7). That is, if A is an $m \times n$ matrix and $0 = 0_{m \times n}$, then

$$A + 0 = 0 + A = A$$

This is the matrix analog of the statement that for any real number a,

$$a + 0 = 0 + a = a$$

With an additive identity in hand, you may ask, "What about a *multiplicative* identity?" In the set of real numbers, the multiplicative identity is the number 1, since

$$a \cdot 1 = 1 \cdot a = a$$

Is there a matrix that plays *this* role? Consider the matrices

$$A = \begin{bmatrix} 1 & 2 \\ 3 & 4 \end{bmatrix} \quad \text{and} \quad I = \begin{bmatrix} 1 & 0 \\ 0 & 1 \end{bmatrix}$$

and verify that

$$AI = \begin{bmatrix} 1 & 2 \\ 3 & 4 \end{bmatrix}\begin{bmatrix} 1 & 0 \\ 0 & 1 \end{bmatrix} = \begin{bmatrix} 1 & 2 \\ 3 & 4 \end{bmatrix} = A$$

and

$$IA = \begin{bmatrix} 1 & 0 \\ 0 & 1 \end{bmatrix}\begin{bmatrix} 1 & 2 \\ 3 & 4 \end{bmatrix} = \begin{bmatrix} 1 & 2 \\ 3 & 4 \end{bmatrix} = A$$

Thus, $AI = IA = A$. In fact, it can be easily shown that for this matrix I, both products AI and IA will equal A for *any* 2×2 matrix A. Therefore,

$$I_2 = \begin{bmatrix} 1 & 0 \\ 0 & 1 \end{bmatrix} = \begin{bmatrix} 1 & \\ & 1 \end{bmatrix}$$

is the multiplicative identity in the set of 2×2 matrices. Similarly, the matrix

$$I_3 = \begin{bmatrix} 1 & 0 & 0 \\ 0 & 1 & 0 \\ 0 & 0 & 1 \end{bmatrix} = \begin{bmatrix} 1 & & \\ & 1 & \\ & & 1 \end{bmatrix}$$

is the multiplicative identity in the set of 3×3 matrices, and so on. (Note that I_3 is the matrix $[\delta_{ij}]_{3\times3}$ encountered in Example

3 above.) In general, the matrix I_n—the $n \times n$ diagonal matrix with every diagonal entry equal to 1—is called the **identity matrix** of order n and serves as the multiplicative identity in the set of all $n \times n$ matrices.

Is there a multiplicative identity in the set of all $m \times n$ matrices if $m \neq n$? For any matrix A in $M_{m \times n}(\mathbf{R})$, the matrix I_m is the **left identity** ($I_m A = A$), and I_n is the **right identity** ($AI_n = A$). Thus, unlike the set of $n \times n$ matrices, the set of nonsquare $m \times n$ matrices does not possess a unique *two-sided* identity, because $I_m \neq I_n$ if $m \neq n$.

Example 20: If A is a square matrix, then A^2 denotes the product AA, A^3 denotes the product AAA, and so forth. If A is the matrix

$$\begin{bmatrix} 0 & -1 \\ 1 & 0 \end{bmatrix}$$

show that $A^3 = -A$.

The calculation

$$A^2 = \begin{bmatrix} 0 & -1 \\ 1 & 0 \end{bmatrix}\begin{bmatrix} 0 & -1 \\ 1 & 0 \end{bmatrix} = \begin{bmatrix} -1 & 0 \\ 0 & -1 \end{bmatrix}$$

shows that $A^2 = -I$. Multiplying both sides of this equation by A yields $A^3 = -A$, as desired. [Technical note: It can be shown that in a certain precise sense, the collection of matrices of the form

$$\begin{bmatrix} a & -b \\ b & a \end{bmatrix}$$

where a and b are real numbers, is structurally identical to the collection of *complex numbers*, $a + bi$. Since the matrix A in this example is of this form (with $a = 0$ and $b = 1$), A corresponds to the complex number $0 + 1i = i$, and the analog of the matrix equation $A^2 = -I$ derived above is $i^2 = -1$, an equation which defines the imaginary unit, i.] ■

Example 21: Find a nondiagonal matrix that commutes with

$$A = \begin{bmatrix} 1 & 2 \\ 3 & 4 \end{bmatrix}$$

The problem is asking for a nondiagonal matrix B such that $AB = BA$. Like A, the matrix B must be 2×2. One way to produce such a matrix B is to form A^2, for if $B = A^2$, associativity implies

$$AB = A \cdot A^2 = A(AA) = (AA)A = A^2 \cdot A = BA$$

(This equation proves that A^2 will commute with A for *any* square matrix A; furthermore, it suggests how one can prove that *every* integral power of a square matrix A will commute with A.)

In this case,

$$B = A^2 = \begin{bmatrix} 1 & 2 \\ 3 & 4 \end{bmatrix}\begin{bmatrix} 1 & 2 \\ 3 & 4 \end{bmatrix} = \begin{bmatrix} 7 & 10 \\ 15 & 22 \end{bmatrix}$$

which is nondiagonal. This matrix B does indeed commute with A, as verified by the calculations

$$AB = \begin{bmatrix} 1 & 2 \\ 3 & 4 \end{bmatrix}\begin{bmatrix} 7 & 10 \\ 15 & 22 \end{bmatrix} = \begin{bmatrix} 37 & 54 \\ 81 & 118 \end{bmatrix}$$

and

$$BA = \begin{bmatrix} 7 & 10 \\ 15 & 22 \end{bmatrix} \begin{bmatrix} 1 & 2 \\ 3 & 4 \end{bmatrix} = \begin{bmatrix} 37 & 54 \\ 81 & 118 \end{bmatrix}$$ ∎

Example 22: If

$$A = \begin{bmatrix} 1 & 1 \\ 0 & 1 \end{bmatrix}$$

prove that

$$A^n = \begin{bmatrix} 1 & n \\ 0 & 1 \end{bmatrix}$$

for every positive integer n.

A few preliminary calculations illustrate that the given formula does hold true:

$$A^2 = \begin{bmatrix} 1 & 1 \\ 0 & 1 \end{bmatrix} \begin{bmatrix} 1 & 1 \\ 0 & 1 \end{bmatrix} = \begin{bmatrix} 1 & 2 \\ 0 & 1 \end{bmatrix}$$

$$A^3 = A^2 \cdot A = \begin{bmatrix} 1 & 2 \\ 0 & 1 \end{bmatrix} \begin{bmatrix} 1 & 1 \\ 0 & 1 \end{bmatrix} = \begin{bmatrix} 1 & 3 \\ 0 & 1 \end{bmatrix}$$

$$A^4 = A^3 \cdot A = \begin{bmatrix} 1 & 3 \\ 0 & 1 \end{bmatrix} \begin{bmatrix} 1 & 1 \\ 0 & 1 \end{bmatrix} = \begin{bmatrix} 1 & 4 \\ 0 & 1 \end{bmatrix}$$

However, to establish that the formula holds for *all* positive integers n, a general proof must be given. This will be done here using *the principle of mathematical induction*, which reads as follows. Let $P(n)$ denote a proposition concerning a positive integer n. If it can be shown that

$$P(1) \text{ is true}$$

and

$$P(n) \text{ is true} \quad \Rightarrow \quad P(n+1) \text{ is true}$$

then the statement $P(n)$ is valid for *all* positive integers n. In the present case, the statement $P(n)$ is the assertion

$$A^n = \begin{bmatrix} 1 & n \\ 0 & 1 \end{bmatrix}$$

Because $A^1 = A$, the statement $P(1)$ is certainly true, since

$$A^1 = \begin{bmatrix} 1 & 1 \\ 0 & 1 \end{bmatrix}$$

Now, assuming that $P(n)$ is true, that is, assuming

$$A^n = \begin{bmatrix} 1 & n \\ 0 & 1 \end{bmatrix}$$

it is now necessary to establish the validity of the statement $P(n+1)$, which is

$$A^{n+1} = \begin{bmatrix} 1 & n+1 \\ 0 & 1 \end{bmatrix}$$

But this statement does indeed hold, because

$$A^{n+1} = A^n \cdot A = \begin{bmatrix} 1 & n \\ 0 & 1 \end{bmatrix}\begin{bmatrix} 1 & 1 \\ 0 & 1 \end{bmatrix} = \begin{bmatrix} 1 & n+1 \\ 0 & 1 \end{bmatrix}$$

By the principle of mathematical induction, the proof is complete. ∎

The inverse of a matrix. Let a be a given real number. Since 1 is the multiplicative identity in the set of real numbers, if a number b exists such that

$$ab = ba = 1$$

then b is called the *reciprocal* or *multiplicative inverse* of a and denoted a^{-1} (or $1/a$). The analog of this statement for square matrices reads as follows. Let A be a given $n \times n$ matrix. Since $I = I_n$ is the multiplicative identity in the set of $n \times n$ matrices, if a matrix B exists such that

$$AB = BA = I$$

then B is called the (multiplicative) **inverse** of A and denoted A^{-1} (read "A inverse").

Example 23: If

$$A = \begin{bmatrix} 2 & 3 \\ 5 & 8 \end{bmatrix}$$

then

$$A^{-1} = \begin{bmatrix} 8 & -3 \\ -5 & 2 \end{bmatrix}$$

since

$$AA^{-1} = \begin{bmatrix} 2 & 3 \\ 5 & 8 \end{bmatrix}\begin{bmatrix} 8 & -3 \\ -5 & 2 \end{bmatrix} = \begin{bmatrix} 1 & 0 \\ 0 & 1 \end{bmatrix} = I$$

and

$$A^{-1}A = \begin{bmatrix} 8 & -3 \\ -5 & 2 \end{bmatrix}\begin{bmatrix} 2 & 3 \\ 5 & 8 \end{bmatrix} = \begin{bmatrix} 1 & 0 \\ 0 & 1 \end{bmatrix} = I \qquad \blacksquare$$

Yet another distinction between the multiplication of scalars and the multiplication of matrices is provided by the existence of inverses. Although every nonzero real number has an inverse, *there exist nonzero matrices that have no inverse.*

Example 24: Show that the nonzero matrix

$$A = \begin{bmatrix} 1 & 1 \\ 0 & 0 \end{bmatrix}$$

has no inverse.

If this matrix had an inverse, then

$$\begin{bmatrix} 1 & 1 \\ 0 & 0 \end{bmatrix}\begin{bmatrix} a & b \\ c & d \end{bmatrix} \quad \text{would equal} \quad \begin{bmatrix} 1 & 0 \\ 0 & 1 \end{bmatrix}$$

for some values of a, b, c, and d. However, since the second row of A is a zero row, you can see that the second row of the product must also be a zero row:

$$\begin{bmatrix} 1 & 1 \\ 0 & 0 \end{bmatrix}\begin{bmatrix} a & b \\ c & d \end{bmatrix} = \begin{bmatrix} * & * \\ 0 & \boxed{0} \end{bmatrix}$$

(When an asterisk, *, appears as an entry in a matrix, it implies that the actual value of this entry is irrelevant to the present discussion.) Since the (2, 2) entry of the product cannot equal 1, the product cannot equal the identity matrix. Therefore, it is impossible to construct a matrix that can serve as the inverse for A. ∎

If a matrix has an inverse, it is said to be **invertible**. The matrix in Example 23 is invertible, but the one in Example 24 is not. Later, you will learn various criteria for determining whether a given square matrix is invertible.

Example 25: Example 23 showed that

$$A = \begin{bmatrix} 2 & 3 \\ 5 & 8 \end{bmatrix} \;\Rightarrow\; A^{-1} = \begin{bmatrix} 8 & -3 \\ -5 & 2 \end{bmatrix}$$

Given that

$$B = \begin{bmatrix} 1 & -2 \\ 2 & -3 \end{bmatrix} \;\Rightarrow\; B^{-1} = \begin{bmatrix} -3 & 2 \\ -2 & 1 \end{bmatrix}$$

verify the equation $(AB)^{-1} = B^{-1}A^{-1}$.

First, compute AB:

$$AB = \begin{bmatrix} 2 & 3 \\ 5 & 8 \end{bmatrix}\begin{bmatrix} 1 & -2 \\ 2 & -3 \end{bmatrix} = \begin{bmatrix} 8 & -13 \\ 21 & -34 \end{bmatrix}$$

Next, compute $B^{-1}A^{-1}$:

$$B^{-1}A^{-1} = \begin{bmatrix} -3 & 2 \\ -2 & 1 \end{bmatrix}\begin{bmatrix} 8 & -3 \\ -5 & 2 \end{bmatrix} = \begin{bmatrix} -34 & 13 \\ -21 & 8 \end{bmatrix}$$

Now, since the product of AB and $B^{-1}A^{-1}$ is I,

$$(AB)(B^{-1}A^{-1}) = \begin{bmatrix} 8 & -13 \\ 21 & -34 \end{bmatrix}\begin{bmatrix} -34 & 13 \\ -21 & 8 \end{bmatrix} = \begin{bmatrix} 1 & 0 \\ 0 & 1 \end{bmatrix} = I$$

$B^{-1}A^{-1}$ is indeed the inverse of AB. In fact, the equation

$$\boxed{(AB)^{-1} = B^{-1}A^{-1}}$$

holds true for *any* invertible square matrices of the same size. This says that if A and B are invertible matrices of the same size, then their product AB is also invertible, and the inverse of the product is equal to the product of the inverses *in the reverse order*. (Compare this equation with the one involving transposes in Example 19 above.) This result can be proved in general by applying the associative law for matrix multiplication. Since

$$(AB)(B^{-1}A^{-1}) = A(BB^{-1})A^{-1} = AIA^{-1} = AA^{-1} = I$$

and

$$(B^{-1}A^{-1})(AB) = B^{-1}(A^{-1}A)B = B^{-1}IB = B^{-1}B = I$$

it follows that $(AB)^{-1} = B^{-1}A^{-1}$, as desired. ∎

Example 26: The inverse of the matrix

$$B = \begin{bmatrix} 1 & -1 & 2 \\ 2 & 0 & 3 \\ 0 & 1 & -1 \end{bmatrix}$$

is

$$B^{-1} = \begin{bmatrix} 3 & -1 & 3 \\ -2 & 1 & -1 \\ -2 & 1 & -2 \end{bmatrix}$$

Show that the inverse of B^{T} is $(B^{-1})^{\mathrm{T}}$.

Form B^T and $(B^{-1})^T$ and multiply:

$$B^T(B^{-1})^T = \begin{bmatrix} 1 & 2 & 0 \\ -1 & 0 & 1 \\ 2 & 3 & -1 \end{bmatrix} \begin{bmatrix} 3 & -2 & -2 \\ -1 & 1 & 1 \\ 3 & -1 & -2 \end{bmatrix} = \begin{bmatrix} 1 & 0 & 0 \\ 0 & 1 & 0 \\ 0 & 0 & 1 \end{bmatrix} = I$$

This calculation shows that $(B^{-1})^T$ is the inverse of B^T. [Strictly speaking, it shows only that $(B^{-1})^T$ is the *right inverse* of B^T, that is, when it multiplies B^T on the right, the product is the identity. It is also true that $(B^{-1})^T B^T = I$, which means $(B^{-1})^T$ is the *left inverse* of B^T. However, it is not necessary to explicitly check both equations: If a square matrix has an inverse, there is no distinction between a left inverse and a right inverse.] Thus,

$$\boxed{(B^T)^{-1} = (B^{-1})^T}$$

an equation which actually holds for *any* invertible square matrix B. This equation says that if a matrix is invertible, then so is its transpose, and the inverse of the transpose is the transpose of the inverse. ∎

Example 27: Use the distributive property for matrix multiplication, $A(B \pm C) = AB \pm AC$, to answer this question: If a 2×2 matrix D satisfies the equation $D^2 - D - 6I = 0$, what is an expression for D^{-1}?

By the distributive property quoted above, $D^2 - D = D^2 - DI = D(D - I)$. Therefore, the equation $D^2 - D - 6I = 0$ implies $D(D - I) = 6I$. Multiplying both sides of this equation by $1/6$ gives

$$D \cdot \left[\tfrac{1}{6}(D - I)\right] = I$$

which implies

$$D^{-1} = \tfrac{1}{6}(D-I)$$

As an illustration of this result, the matrix

$$D = \begin{bmatrix} 4 & -2 \\ 3 & -3 \end{bmatrix}$$

satisfies the equation $D^2 - D - 6I = 0$, as you may verify. Since

$$\tfrac{1}{6}(D-I) = \tfrac{1}{6}\left(\begin{bmatrix} 4 & -2 \\ 3 & -3 \end{bmatrix} - \begin{bmatrix} 1 & 0 \\ 0 & 1 \end{bmatrix} \right) = \tfrac{1}{6}\begin{bmatrix} 3 & -2 \\ 3 & -4 \end{bmatrix} = \begin{bmatrix} \tfrac{1}{2} & -\tfrac{1}{3} \\ \tfrac{1}{2} & -\tfrac{2}{3} \end{bmatrix}$$

and

$$D \cdot \left[\tfrac{1}{6}(D-I) \right] = \begin{bmatrix} 4 & -2 \\ 3 & -3 \end{bmatrix} \begin{bmatrix} \tfrac{1}{2} & -\tfrac{1}{3} \\ \tfrac{1}{2} & -\tfrac{2}{3} \end{bmatrix} = \begin{bmatrix} 1 & 0 \\ 0 & 1 \end{bmatrix} = I$$

the matrix $\tfrac{1}{6}(D-I)$ does indeed equal D^{-1}, as claimed. ∎

Example 28: The equation $(a + b)^2 = a^2 + 2ab + b^2$ is an identity if a and b are real numbers. Show, however, that $(A + B)^2 = A^2 + 2AB + B^2$ is *not* an identity if A and B are 2×2 matrices. [Note: The distributive laws for matrix multiplication are $A(B \pm C) = AB \pm AC$, given in Example 27, and the companion law, $(A \pm B)C = AC \pm BC$.]

The distributive laws for matrix multiplication imply

$$(A+B)^2 \equiv (A+B)(A+B)$$
$$= (A+B)A + (A+B)B$$
$$= (AA+BA) + (AB+BB)$$
$$= A^2 + BA + AB + B^2$$

Since matrix multiplication is not commutative, BA will usually not equal AB, so the sum $BA + AB$ cannot be written as $2AB$. In general, then, $(A+B)^2 \neq A^2 + 2AB + B^2$. [Any matrices A and B that do not commute (for example, the matrices in Example 16 above) would provide a specific counterexample to the statement $(A+B)^2 = A^2 + 2AB + B^2$, which would also establish that this is not an identity.] ∎

Example 29: Assume that B is invertible. If A commutes with B, show that A will also commute with B^{-1}.

Proof. To say "A commutes with B" means $AB = BA$. Multiply this equation by B^{-1} on the left and on the right and use associativity:

$$B^{-1}(AB)B^{-1} = B^{-1}(BA)B^{-1}$$
$$(B^{-1}A)(BB^{-1}) = (B^{-1}B)(AB^{-1})$$
$$B^{-1}A = AB^{-1} \quad \blacksquare$$

Example 30: The number 0 has just one square root: 0. Show, however, that the (2 by 2) zero matrix has infinitely many square roots by finding all 2×2 matrices A such that $A^2 = 0$.

In the same way that a number a is called a square root of b if $a^2 = b$, a matrix A is said to be a square root of B if $A^2 = B$. Let

$$A = \begin{bmatrix} a & b \\ c & d \end{bmatrix}$$

be an arbitrary 2×2 matrix. Squaring it and setting the result equal to 0 gives

$$A^2 = \begin{bmatrix} a & b \\ c & d \end{bmatrix}\begin{bmatrix} a & b \\ c & d \end{bmatrix} = \begin{bmatrix} a^2 + bc & b(a+d) \\ c(a+d) & bc + d^2 \end{bmatrix} \overset{\text{set}}{=} \begin{bmatrix} 0 & 0 \\ 0 & 0 \end{bmatrix}$$

The $(1, 2)$ entries in the last equation imply $b(a + d) = 0$, which holds if (Case 1) $b = 0$ or (Case 2) $d = -a$.

• Case 1. If $b = 0$, the diagonal entries then imply $a = 0$ and $d = 0$, and the $(2, 1)$ entries imply that c is arbitrary. Thus, for any value of c, every matrix of the form

$$\begin{bmatrix} 0 & 0 \\ c & 0 \end{bmatrix}$$

is a square root of $0_{2\times2}$.

• Case 2. If $d = -a$, then the off-diagonal entries will both be 0, and the diagonal entries will both equal $a^2 + bc$. Thus, as long as b and c are chosen so that $bc = -a^2$, A^2 will equal 0.

A similar chain of reasoning beginning with the $(2, 1)$ entries leads to either $a = c = d = 0$ (and b arbitrary) or the same conclusion as before: as long as b and c are chosen so that $bc = -a^2$, the matrix A^2 will equal 0.

All these cases can be summarized as follows. Any matrix of the following form will have the property that its square is the 2 by 2 zero matrix:

$$\begin{bmatrix} a & b \\ c & -a \end{bmatrix}, \quad \text{with } bc = -a^2$$

Since there are infinitely many values of a, b, and c such that $bc = -a^2$, the zero matrix $0_{2\times 2}$ has infinitely many square roots. For example, choosing $a = 4$, $b = 2$, and $c = -8$ gives the nonzero matrix

$$S = \begin{bmatrix} 4 & 2 \\ -8 & -4 \end{bmatrix}$$

whose square is

$$S^2 = \begin{bmatrix} 4 & 2 \\ -8 & -4 \end{bmatrix} \begin{bmatrix} 4 & 2 \\ -8 & -4 \end{bmatrix} = \begin{bmatrix} 0 & 0 \\ 0 & 0 \end{bmatrix} = 0 \quad \blacksquare$$

The basic problem of linear algebra is to solve a system of linear equations. A **linear** equation in the n variables—or unknowns—$x_1, x_2, \ldots$, and x_n is an equation of the form

$$a_1x_1 + a_2x_2 + \cdots + a_nx_n = b$$

where b and the coefficients a_i are constants. A finite collection of such linear equations is called a **linear system**. To **solve a system** means to find all values of the variables that satisfy all the equations in the system *simultaneously*. For example, consider the following system, which consists of two linear equations in two unknowns:

$$x_1 + x_2 = 3$$
$$3x_1 - 2x_2 = 4$$

Although there are infinitely many solutions to each equation separately, there is only one pair of numbers x_1 and x_2 which satisfies both equations at the same time. This ordered pair, $(x_1, x_2) = (2, 1)$, is called the **solution** to the system.

Solutions to Linear Systems

The analysis of linear systems will begin by determining the possibilities for the solutions. Despite the fact that the system can contain any number of equations, each of which can involve any number of unknowns, the result that describes the possible number of solutions to a linear system is simple and definitive. The fundamental ideas will be illustrated in the following examples.

Example 1: Interpret the following system graphically:

$$x + y = 3$$
$$3x - 2y = 4$$

Each of these equations specifies a line in the x-y plane, and every point on each line represents a solution to its equation. Therefore, the point where the lines cross—(2, 1)—satisfies both equations simultaneously; this is the solution to the system. See Figure 36.

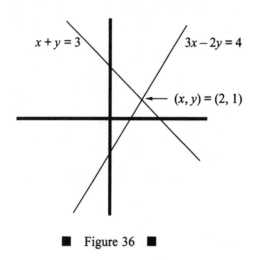

$x + y = 3$ $3x - 2y = 4$

$(x, y) = (2, 1)$

■ Figure 36 ■

Example 2: Interpret this system graphically:

$$x + y = 3$$
$$x + y = -2$$

The lines specified by these equations are parallel and do not intersect, as shown in Figure 37. Since there is no point of intersection, there is no solution to this system. (Clearly, the sum of two numbers cannot be both 3 and –2.) A system which has no solutions—such as this one—is said to be **inconsistent**.

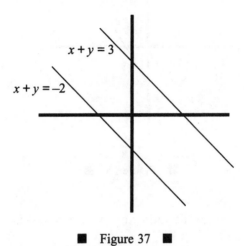

■ Figure 37 ■

Example 3: Interpret the following system graphically:

$$3x - 2y = 4$$
$$6x - 4y = 8$$

Since the second equation is merely a constant multiple of the first, the lines specified by these equations are identical, as shown in Figure 38. Clearly then, every solution to the first equation is automatically a solution to the second as well, so this system has infinitely many solutions.

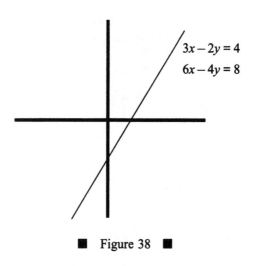

$$3x - 2y = 4$$
$$6x - 4y = 8$$

■ Figure 38 ■

Example 4: Discuss the following system graphically:

$$x - 2y + z = 0$$
$$2x + y - 3z = -5$$

Each of these equations specifies a plane in $\mathbf{R}^3$. Two such planes either coincide, intersect in a line, or are distinct and parallel. Therefore, a system of two equations in three unknowns has either no solutions or infinitely many. For this particular system, the planes do not coincide, as can be seen, for example, by noting that the first plane passes through the origin while the second does not. These planes are not parallel, since $\mathbf{v}_1 = (1, -2, 1)$ is normal to the first and $\mathbf{v}_2 = (2, 1, -3)$ is normal to the second, and neither of these vectors is a scalar multiple of the other. Therefore, these planes intersect in a line, and the system has infinitely many solutions.   ■

Example 5: Interpret the following system graphically:

$$x + y = 3$$
$$3x - 2y = 4$$
$$x + 3y = 9$$

Each of these equations specifies a line in the x-y plane, as sketched in Figure 39. Note that while any *two* of these lines have a point of intersection, there is no point common to all *three* lines. This system is inconsistent.

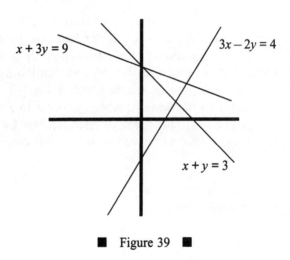

$x + 3y = 9$ $3x - 2y = 4$ $x + y = 3$

■ Figure 39 ■

These examples illustrate the three possibilities for the solutions to a linear system:

Theorem A. Regardless of its size or the number of unknowns its equations contain, a linear system will have either no solutions, exactly one solution, or infinitely many solutions.

This will be proved in Example 18 below. Example 4 illustrated the following additional fact about the solutions to a linear system:

Theorem B. If there are fewer equations than unknowns, then the system will have either no solutions or infinitely many.

Gaussian Elimination

The purpose of this section is to describe how the solutions to a linear system are actually found. The fundamental idea is to add multiples of one equation to the others in order to eliminate a variable and to continue this process until only one variable is left. Once this final variable is determined, its value is substituted back into the other equations in order to evaluate the remaining unknowns. This method, characterized by step-by-step elimination of the variables, is called **Gaussian elimination**.

Example 6: Solve this system:

$$x + y = 3$$
$$3x - 2y = 4$$

Multiplying the first equation by −3 and adding the result to the second equation eliminates the variable x:

$$
\begin{array}{r}
-3x - 3y = -9 \\
3x - 2y = 4 \\
\hline
-5y = -5
\end{array}
$$

This final equation, $-5y = -5$, immediately implies $y = 1$. Back-substitution of $y = 1$ into the original first equation, $x + y = 3$, yields $x = 2$. (Back-substitution of $y = 1$ into the original second equation, $3x - 2y = 4$, would also yield $x = 2$.) The solution of this system is therefore $(x, y) = (2, 1)$, as noted in Example 1. ■

Gaussian elimination is usually carried out using matrices. This method reduces the effort in finding the solutions by eliminating the need to explicitly write the variables at each step. The previous example will be redone using matrices.

Example 7: Solve this system:

$$x + y = 3$$
$$3x - 2y = 4$$

The first step is to write the coefficients of the unknowns in a matrix:

$$\begin{bmatrix} 1 & 1 \\ 3 & -2 \end{bmatrix}$$

This is called the **coefficient matrix** of the system. Next, the coefficient matrix is augmented by writing the constants that appear on the right-hand sides of the equations as an additional column:

$$\left[\begin{array}{cc|c} 1 & 1 & 3 \\ 3 & -2 & 4 \end{array}\right]$$

This is called the **augmented matrix**, and each row corresponds to an equation in the given system. The first row, $\mathbf{r}_1 =$

(1, 1, 3), corresponds to the first equation, $1x + 1y = 3$, and the second row, $r_2 = (3, -2, 4)$, corresponds to the second equation, $3x - 2y = 4$. You may choose to include a vertical line—as shown above—to separate the coefficients of the unknowns from the extra column representing the constants.

Now, the counterpart of eliminating a variable from an equation in the system is changing one of the entries in the coefficient matrix to zero. Likewise, the counterpart of adding a multiple of one equation to another is adding a multiple of one row to another row. Adding -3 times the first row of the augmented matrix to the second row yields

$$\begin{bmatrix} 1 & 1 & 3 \\ 3 & -2 & 4 \end{bmatrix} \xrightarrow{\ -3r_1 \text{ added to } r_2\ } \begin{bmatrix} 1 & 1 & 3 \\ 0 & -5 & -5 \end{bmatrix}$$

The new second row translates into $-5y = -5$, which means $y = 1$. Back-substitution into the first row (that is, into the equation that represents the first row) yields $x = 2$ and, therefore, the solution to the system: $(x, y) = (2, 1)$. ∎

Gaussian elimination can be summarized as follows. Given a linear system expressed in matrix form, $A\mathbf{x} = \mathbf{b}$, first write down the corresponding augmented matrix:

$$\begin{bmatrix} A \,|\, \mathbf{b} \end{bmatrix}$$

Then, perform a sequence of **elementary row operations**, which are any of the following:

Type 1. Interchange any two rows.
Type 2. Multiply a row by a nonzero constant.
Type 3. Add a multiple of one row to another row.

The goal of these operations is to transform—or **reduce**—the original augmented matrix into one of the form

$$[A' | \mathbf{b}']$$

where A' is upper triangular ($a'_{ij} = 0$ for $i > j$), any zero rows appear at the bottom of the matrix, and the first nonzero entry in any row is to the right of the first nonzero entry in any higher row; such a matrix is said to be in **echelon** form. The solutions of the system represented by the simpler augmented matrix, $[A' | \mathbf{b}']$, can be found by inspection of the bottom rows and back-substitution into the higher rows. Since elementary row operations do not change the solutions of the system, the vectors $\mathbf{x}$ which satisfy the simpler system $A'\mathbf{x} = \mathbf{b}'$ are precisely those that satisfy the original system, $A\mathbf{x} = \mathbf{b}$.

Example 8: Solve the following system using Gaussian elimination:

$$\begin{array}{rcrcrcr} x & - & 2y & + & z & = & 0 \\ 2x & + & y & - & 3z & = & 5 \\ 4x & - & 7y & + & z & = & -1 \end{array}$$

The augmented matrix which represents this system is

$$\left[\begin{array}{ccc|c} 1 & -2 & 1 & 0 \\ 2 & 1 & -3 & 5 \\ 4 & -7 & 1 & -1 \end{array}\right]$$

The first goal is to produce zeros below the first entry in the first column, which translates into eliminating the first variable, x, from the second and third equations. The row operations which accomplish this are as follows:

$$\begin{bmatrix} 1 & -2 & 1 & 0 \\ 2 & 1 & -3 & 5 \\ 4 & -7 & 1 & -1 \end{bmatrix} \xrightarrow[\substack{-4r_1 \text{ added to } r_3}]{\substack{-2r_1 \text{ added to } r_2}} \begin{bmatrix} 1 & -2 & 1 & 0 \\ 0 & 5 & -5 & 5 \\ 0 & 1 & -3 & -1 \end{bmatrix}$$

The second goal is to produce a zero below the second entry in the second column, which translates into eliminating the second variable, y, from the third equation. One way to accomplish this would be to add $-1/5$ times the second row to the third row. However, to avoid fractions, there is another option: first interchange rows two and three. Interchanging two rows merely interchanges the equations, which clearly will not alter the solution of the system:

$$\begin{bmatrix} 1 & -2 & 1 & 0 \\ 0 & 5 & -5 & 5 \\ 0 & 1 & -3 & -1 \end{bmatrix} \xrightarrow{r_2 \leftrightarrow r_3} \begin{bmatrix} 1 & -2 & 1 & 0 \\ 0 & 1 & -3 & -1 \\ 0 & 5 & -5 & 5 \end{bmatrix}$$

Now, add -5 times the second row to the third row:

$$\begin{bmatrix} 1 & -2 & 1 & 0 \\ 0 & 1 & -3 & -1 \\ 0 & 5 & -5 & 5 \end{bmatrix} \xrightarrow{-5r_2 \text{ added to } r_3} \begin{bmatrix} 1 & -2 & 1 & 0 \\ 0 & 1 & -3 & -1 \\ \underbrace{0 \quad 0 \quad 10 \quad 10}_{\text{echelon form}} \end{bmatrix}$$

Since the coefficient matrix has been transformed into echelon form, the "forward" part of Gaussian elimination is complete. What remains now is to use the third row to evaluate the third unknown, then to back-substitute into the second row to evaluate the second unknown, and, finally, to back-substitute into the first row to evaluate the first unknown.

The third row of the final matrix translates into $10z = 10$, which gives $z = 1$. Back-substitution of this value into the second row, which represents the equation $y - 3z = -1$, yields $y = 2$. Back-substitution of both these values into the first row, which represents the equation $x - 2y + z = 0$, gives $x = 3$. The solution of this system is therefore $(x, y, z) = (3, 2, 1)$. ∎

Example 9: Solve the following system using Gaussian elimination:

$$\begin{aligned} 2x &- 2y && = -6 \\ x &- y &+ z &= 1 \\ &3y &- 2z &= -5 \end{aligned}$$

For this system, the augmented matrix (vertical line omitted) is

$$\begin{bmatrix} 2 & -2 & 0 & -6 \\ 1 & -1 & 1 & 1 \\ 0 & 3 & -2 & -5 \end{bmatrix}$$

First, multiply row 1 by $\frac{1}{2}$:

$$\begin{bmatrix} 2 & -2 & 0 & -6 \\ 1 & -1 & 1 & 1 \\ 0 & 3 & -2 & -5 \end{bmatrix} \xrightarrow{\text{Multiply } r_1 \text{ by } \frac{1}{2}} \begin{bmatrix} 1 & -1 & 0 & -3 \\ 1 & -1 & 1 & 1 \\ 0 & 3 & -2 & -5 \end{bmatrix}$$

Now, adding -1 times the first row to the second row yields zeros below the first entry in the first column:

$$\begin{bmatrix} 1 & -1 & 0 & -3 \\ 1 & -1 & 1 & 1 \\ 0 & 3 & -2 & -5 \end{bmatrix} \xrightarrow{-r_1 \text{ added ro } r_2} \begin{bmatrix} 1 & -1 & 0 & -3 \\ 0 & 0 & 1 & 4 \\ 0 & 3 & -2 & -5 \end{bmatrix}$$

Interchanging the second and third rows then gives the desired upper-triangular coefficient matrix:

$$\begin{bmatrix} 1 & -1 & 0 & -3 \\ 0 & 0 & 1 & 4 \\ 0 & 3 & -2 & -5 \end{bmatrix} \xrightarrow{\;r_2 \leftrightarrow r_3\;} \begin{bmatrix} 1 & -1 & 0 & -3 \\ 0 & 3 & -2 & -5 \\ 0 & 0 & 1 & 4 \end{bmatrix}$$

The third row now says $z = 4$. Back-substituting this value into the second row gives $y = 1$, and back-substitution of both these values into the first row yields $x = -2$. The solution of this system is therefore $(x, y, z) = (-2, 1, 4)$. ■

Gauss-Jordan elimination. Gaussian elimination proceeds by performing elementary row operations to produce zeros below the diagonal of the coefficient matrix to reduce it to echelon form. (Recall that a matrix $A' = [a'_{ij}]$ is in echelon form when $a'_{ij} = 0$ for $i > j$, any zero rows appear at the bottom of the matrix, and the first nonzero entry in any row is to the right of the first nonzero entry in any higher row.) Once this is done, inspection of the bottom row(s) and back-substitution into the upper rows determine the values of the unknowns.

However, it is possible to reduce (or eliminate entirely) the computations involved in back-substitution by performing additional row operations to transform the matrix from echelon form to **reduced echelon** form. A matrix is in reduced echelon form when, in addition to being in echelon form, each column that contains a nonzero entry (usually made to be 1) has zeros not just below that entry but also above that entry. Loosely speaking, Gaussian elimination works from the top down, to produce a matrix in echelon form, whereas **Gauss-Jordan elimination** continues where Gaussian left off by then working from the bottom up to produce a matrix in reduced echelon form. The technique will be illustrated in the following example.

Example 10: The height, y, of an object thrown into the air is known to be given by a quadratic function of t (time) of the form $y = at^2 + bt + c$. If the object is at height $y = 23/4$ at time $t = 1/2$, at $y = 7$ at time $t = 1$, and at $y = 2$ at $t = 2$, determine the coefficients a, b, and c.

Since $t = 1/2$ gives $y = 23/4$,

$$\tfrac{23}{4} = a(\tfrac{1}{2})^2 + b(\tfrac{1}{2}) + c$$
$$= \tfrac{1}{4}a + \tfrac{1}{2}b + c$$

while the other two conditions, $y(t = 1) = 7$ and $y(t = 2) = 2$, give the following equations for a, b, and c:

$$7 = a + b + c$$
$$2 = 4a + 2b + c$$

Therefore, the goal is solve the system

$$\tfrac{1}{4}a + \tfrac{1}{2}b + c = \tfrac{23}{4}$$
$$a + b + c = 7$$
$$4a + 2b + c = 2$$

The augmented matrix for this system is reduced as follows:

$$\begin{bmatrix} \frac{1}{4} & \frac{1}{2} & 1 & \frac{23}{4} \\ 1 & 1 & 1 & 7 \\ 4 & 2 & 1 & 2 \end{bmatrix} \xrightarrow{4r_1} \begin{bmatrix} 1 & 2 & 4 & 23 \\ 1 & 1 & 1 & 7 \\ 4 & 2 & 1 & 2 \end{bmatrix}$$

$$\xrightarrow[-4r_1 \text{ added to } r_3]{-r_1 \text{ added to } r_2} \begin{bmatrix} 1 & 2 & 4 & 23 \\ 0 & -1 & -3 & -16 \\ 0 & -6 & -15 & -90 \end{bmatrix}$$

$$\xrightarrow[-r_2]{-6r_2 \text{ added to } r_3} \begin{bmatrix} 1 & 2 & 4 & 23 \\ 0 & 1 & 3 & 16 \\ 0 & 0 & 3 & 6 \end{bmatrix}$$

At this point, the forward part of Gaussian elimination is finished, since the coefficient matrix has been reduced to echelon form. However, to illustrate Gauss-Jordan elimination, the following additional elementary row operations are performed:

$$\begin{bmatrix} 1 & 2 & 4 & 23 \\ 0 & 1 & 3 & 16 \\ 0 & 0 & 3 & 6 \end{bmatrix} \xrightarrow[\frac{1}{3}r_3]{-r_3 \text{ added to } r_2} \begin{bmatrix} 1 & 2 & 4 & 23 \\ 0 & 1 & 0 & 10 \\ 0 & 0 & 1 & 2 \end{bmatrix}$$

$$\xrightarrow{-4r_3 \text{ added to } r_1} \begin{bmatrix} 1 & 2 & 0 & 15 \\ 0 & 1 & 0 & 10 \\ 0 & 0 & 1 & 2 \end{bmatrix}$$

$$\xrightarrow{-2r_2 \text{ added to } r_1} \begin{bmatrix} 1 & 0 & 0 & -5 \\ 0 & 1 & 0 & 10 \\ 0 & 0 & 1 & 2 \end{bmatrix}$$

This final matrix immediately gives the solution: $a = -5$, $b = 10$, and $c = 2$. ∎

Example 11: Solve the following system using Gaussian elimination:

$$
\begin{aligned}
x + y - 3z &= 4 \\
2x + y - z &= 2 \\
3x + 2y - 4z &= 7
\end{aligned}
$$

The augmented matrix for this system is

$$
\begin{bmatrix}
1 & 1 & -3 & 4 \\
2 & 1 & -1 & 2 \\
3 & 2 & -4 & 7
\end{bmatrix}
$$

Multiples of the first row are added to the other rows to produce zeros below the first entry in the first column:

$$
\begin{bmatrix}
1 & 1 & -3 & 4 \\
2 & 1 & -1 & 2 \\
3 & 2 & -4 & 7
\end{bmatrix}
\xrightarrow[-3r_1 \text{ added to } r_3]{-2r_1 \text{ added to } r_2}
\begin{bmatrix}
1 & 1 & -3 & 4 \\
0 & -1 & 5 & -6 \\
0 & -1 & 5 & -5
\end{bmatrix}
$$

Next, -1 times the second row is added to the third row:

$$
\begin{bmatrix}
1 & 1 & -3 & 4 \\
0 & -1 & 5 & -6 \\
0 & -1 & 5 & -5
\end{bmatrix}
\xrightarrow{-r_2 \text{ added to } r_3}
\begin{bmatrix}
1 & 1 & -3 & 4 \\
0 & -1 & 5 & -6 \\
0 & 0 & 0 & 1
\end{bmatrix}
$$

The third row now says $0x + 0y + 0z = 1$, an equation that cannot be satisfied by any values of x, y, and z. The process stops: this system has no solutions. ∎

The previous example shows how Gaussian elimination reveals an inconsistent system. A slight alteration of that system (for example, changing the constant term "7" in the third equation to a "6") will illustrate a system with infinitely many solutions.

Example 12: Solve the following system using Gaussian elimination:

$$x + y - 3z = 4$$
$$2x + y - z = 2$$
$$3x + 2y - 4z = 6$$

The same operations applied to the augmented matrix of the system in Example 11 are applied to the augmented matrix for the present system:

$$\begin{bmatrix} 1 & 1 & -3 & 4 \\ 2 & 1 & -1 & 2 \\ 3 & 2 & -4 & 6 \end{bmatrix} \xrightarrow[\substack{-2r_1 \text{ added to } r_2 \\ -3r_1 \text{ added to } r_3}]{} \begin{bmatrix} 1 & 1 & -3 & 4 \\ 0 & -1 & 5 & -6 \\ 0 & -1 & 5 & -6 \end{bmatrix}$$

$$\xrightarrow[-r_2 \text{ added to } r_3]{} \begin{bmatrix} 1 & 1 & -3 & 4 \\ 0 & -1 & 5 & -6 \\ 0 & 0 & 0 & 0 \end{bmatrix}$$

Here, the third row translates into $0x + 0y + 0z = 0$, an equation which is satisfied by *any* x, y, and z. Since this offers no constraint on the unknowns, there are not three conditions on the unknowns, only two (represented by the two nonzero rows in the final augmented matrix). Since there are 3 unknowns but only 2 constraints, $3 - 2 = 1$ of the unknowns, z say, is arbitrary; this is called a **free variable**. Let $z = t$, where t is any real number. Back-substitution of $z = t$ into the second row $(-y + 5z = -6)$ gives

$$-y + 5t = -6 \quad \Rightarrow \quad y = 6 + 5t$$

Back substituting $z = t$ and $y = 6 + 5t$ into the first row ($x + y - 3z = 4$) determines x:

$$x + (6 + 5t) - 3t = 4 \quad \Rightarrow \quad x = -2 - 2t$$

Therefore, every solution of the system has the form

$$(x, y, z) = (-2 - 2t, 6 + 5t, t) = (-2t, 5t, t) + (-2, 6, 0) \quad (*)$$

where t is any real number. There are infinitely many solutions, since every real value of t gives a different particular solution. For example, choosing $t = 1$ gives $(x, y, z) = (-4, 11, 1)$, while $t = -3$ gives $(x, y, z) = (4, -9, -3)$, and so on. Geometrically, this system represents three planes in $\mathbf{R}^3$ that intersect in a line, and (*) is a parametric equation for this line. ∎

Example 12 provided an illustration of a system with infinitely many solutions, how this case arises, and how the solution is written. Every linear system that possesses infinitely many solutions must contain at least one arbitrary **parameter** (free variable). Once the augmented matrix has been reduced to echelon form, the number of free variables is equal to the total number of unknowns minus the number of nonzero rows:

> # free variables
> = # unknowns − # nonzero rows in echelon form

This agrees with Theorem B above, which states that a linear system with fewer equations than unknowns, if consistent, has infinitely many solutions. The condition "fewer equations than unknowns" means that the number of rows in the coefficient matrix is less than the number of unknowns. Therefore, the boxed equation above implies that there must be at least one free variable. Since such a variable can, by definition, take on infinitely many values, the system will have infinitely many solutions.

Example 13: Find all solutions to the system

$$
\begin{aligned}
w - x + y - z &= 1 \\
2w + x - 3y &= 2 \\
5w - 2x - 3z &= 5
\end{aligned}
$$

First, note that there are four unknowns, but only three equations. Therefore, if the system is consistent, it is guaranteed to have infinitely many solutions, a condition characterized by at least one parameter in the general solution. After the corresponding augmented matrix is constructed, Gaussian elimination yields

$$
\begin{bmatrix}
1 & -1 & 1 & -1 & 1 \\
2 & 1 & -3 & 0 & 2 \\
5 & -2 & 0 & -3 & 5
\end{bmatrix}
\xrightarrow[\substack{-5r_1 \text{ added to } r_3}]{-2r_1 \text{ added to } r_2}
\begin{bmatrix}
1 & -1 & 1 & -1 & 1 \\
0 & 3 & -5 & 2 & 0 \\
0 & 3 & -5 & 2 & 0
\end{bmatrix}
$$

$$
\xrightarrow{-r_2 \text{ added to } r_3}
\begin{bmatrix}
1 & -1 & 1 & -1 & 1 \\
0 & 3 & -5 & 2 & 0 \\
0 & 0 & 0 & 0 & 0
\end{bmatrix}
$$

The fact that only two nonzero rows remain in the echelon form of the augmented matrix means that $4 - 2 = 2$ of the variables are free:

free variables = # unknowns − # nonzero rows in echelon form
$$= 4 - 2$$
$$= 2$$

Therefore, selecting y and z as the free variables, let $y = t_1$ and $z = t_2$. The second row of the reduced augmented matrix implies

$$3x - 5t_1 + 2t_2 = 0 \quad \Rightarrow \quad x = \tfrac{1}{3}(5t_1 - 2t_2)$$

and the first row then gives

$$w - \tfrac{1}{3}(5t_1 - 2t_2) + t_1 - t_2 = 1 \quad \Rightarrow \quad w = 1 + \tfrac{1}{3}(2t_1 + t_2)$$

Thus, the solutions of the system have the form

$$(w, x, y, z) = \left(1 + \tfrac{1}{3}(2t_1 + t_2), \ \tfrac{1}{3}(5t_1 - 2t_2), \ t_1, \ t_2\right)$$

where t_1 and t_2 are allowed to take on any real values. ∎

Example 14: Let $\mathbf{b} = (b_1, b_2, b_3)^{\mathrm{T}}$ and let A be the matrix

$$\begin{bmatrix} 2 & 1 & -1 \\ -1 & -3 & 1 \\ 1 & 8 & -2 \end{bmatrix}$$

For what values of b_1, b_2, and b_3 will the system $A\mathbf{x} = \mathbf{b}$ be consistent?

The augmented matrix for the system $A\mathbf{x} = \mathbf{b}$ reads

$$\begin{bmatrix} 2 & 1 & -1 & b_1 \\ -1 & -3 & 1 & b_2 \\ 1 & 8 & -2 & b_3 \end{bmatrix}$$

which Gaussian elimination reduces as follows:

$$\begin{bmatrix} 2 & 1 & -1 & b_1 \\ -1 & -3 & 1 & b_2 \\ 1 & 8 & -2 & b_3 \end{bmatrix} \xrightarrow{\mathbf{r}_1 \leftrightarrow \mathbf{r}_2} \begin{bmatrix} -1 & -3 & 1 & b_2 \\ 2 & 1 & -1 & b_1 \\ 1 & 8 & -2 & b_3 \end{bmatrix}$$

$$\xrightarrow[\substack{2\mathbf{r}_1 \text{ added to } \mathbf{r}_2 \\ \mathbf{r}_1 \text{ added to } \mathbf{r}_3}]{} \begin{bmatrix} -1 & -3 & 1 & b_2 \\ 0 & -5 & 1 & b_1 + 2b_2 \\ 0 & 5 & -1 & b_2 + b_3 \end{bmatrix}$$

$$\xrightarrow{\mathbf{r}_2 \text{ added to } \mathbf{r}_3} \begin{bmatrix} -1 & -3 & 1 & b_2 \\ 0 & -5 & 1 & b_1 + 2b_2 \\ 0 & 0 & 0 & b_1 + 3b_2 + b_3 \end{bmatrix}$$

The bottom row now implies that $b_1 + 3b_2 + b_3$ must be zero if this system is to be consistent. Therefore, the given system has solutions (infinitely many, in fact) only for those column vectors $\mathbf{b} = (b_1, b_2, b_3)^T$ for which $b_1 + 3b_2 + b_3 = 0$. ∎

Example 15: Solve the following system (compare to Example 12):

$$\begin{aligned} x + y - 3z &= 0 \\ 2x + y - z &= 0 \\ 3x + 2y - 4z &= 0 \end{aligned}$$

A system such as this one, where the constant term on the right-hand side of *every* equation is 0, is called a **homogeneous** system. In matrix form it reads $A\mathbf{x} = \mathbf{0}$. Since every homogeneous system is consistent—because $\mathbf{x} = \mathbf{0}$ is always a solution—a homogeneous system has either exactly one solution (the **trivial** solution, $\mathbf{x} = \mathbf{0}$) or infinitely many. The row-

LINEAR
SYSTEMS

reduction of the coefficient matrix for this system has already
been performed in Example 12. It is not necessary to explicitly augment the coefficient matrix with the column $\mathbf{b} = \mathbf{0}$,
since no elementary row operation can affect these zeros. That
is, if A' is an echelon form of A, then elementary row opera-
tions will transform $[A \mid \mathbf{0}]$ into $[A' \mid \mathbf{0}]$. From the result of Example 12,

$$[A' \mid \mathbf{0}] = \begin{bmatrix} 1 & 1 & -3 & 0 \\ 0 & -1 & 5 & 0 \\ 0 & 0 & 0 & 0 \end{bmatrix}$$

Since the last row again implies that z can be taken as a free
variable, let $z = t$, where t is any real number. Back-substitution
of $z = t$ into the second row $(-y + 5z = 0)$ gives

$$-y + 5t = 0 \quad \Rightarrow \quad y = 5t$$

and back-substitution of $z = t$ and $y = 5t$ into the first row $(x +
y - 3z = 0)$ determines x:

$$x + 5t - 3t = 0 \quad \Rightarrow \quad x = -2t$$

Therefore, every solution of this system has the form $(x, y, z) =
(-2t, 5t, t)$, where t is any real number. There are infinitely
many solutions, since every real value of t gives a unique particular solution.

Note carefully the difference between the set of solutions
to the system in Example 12 and the one here. Although both
had the same coefficient matrix A, the system in Example 12
was nonhomogeneous $(A\mathbf{x} = \mathbf{b}$, where $\mathbf{b} \neq \mathbf{0})$, while the one
here is the corresponding homogeneous system, $A\mathbf{x} = \mathbf{0}$.
Placing their solutions side by side,

general solution to $A\mathbf{x} = \mathbf{0}$: $(x, y, z) = (-2t, 5t, t)$

general solution to $A\mathbf{x} = \mathbf{b}$: $(x, y, z) = (-2t, 5t, t) + (-2, 6, 0)$

illustrates an important fact:

Theorem C. The general solution to a consistent nonhomogeneous linear system, $A\mathbf{x} = \mathbf{b}$, is equal to the general solution of the corresponding homogeneous system, $A\mathbf{x} = \mathbf{0}$, plus a particular solution of the nonhomogeneous system. That is, if $\mathbf{x} = \mathbf{x}_h$ represents the general solution of $A\mathbf{x} = \mathbf{0}$, then $\mathbf{x} = \mathbf{x}_h + \overline{\mathbf{x}}$ represents the general solution of $A\mathbf{x} = \mathbf{b}$, where $\overline{\mathbf{x}}$ is any particular solution of the (consistent) nonhomogeneous system $A\mathbf{x} = \mathbf{b}$.

[Technical note: Theorem C, which concerns a *linear system*, has a counterpart in the theory of *linear differential equations*. Let L be a linear differential operator; then the general solution of a solvable nonhomogeneous linear differential equation, $L(y) = d$ (where $d \not\equiv 0$), is equal to the general solution of the corresponding homogeneous equation, $L(y) = 0$, plus a particular solution of the nonhomogeneous equation. That is, if $y = y_h$ represents the general solution of $L(y) = 0$, then $y = y_h + \overline{y}$ represents the general solution of $L(y) = d$, where $\overline{y}$ is any particular solution of the (solvable) nonhomogeneous linear equation $L(y) = d$. (This is Theorem B on page 68 in the present author's *Differential Equations*, © 1995 Cliffs Notes, Inc.)] ■

Example 16: Determine all solutions of the system

$$
\begin{array}{rrrrrrr}
 & x & - & 3y & + & 4z & = & 1 \\
2w & - & 2x & + & y & & = & -1 \\
2w & - & x & - & 2y & + & 4z & = & 0 \\
-6w & + & 4x & + & 3y & - & 8z & = & 1
\end{array}
$$

Write down the augmented matrix and perform the following sequence of operations:

$$
\begin{bmatrix}
0 & 1 & -3 & 4 & 1 \\
2 & -2 & 1 & 0 & -1 \\
2 & -1 & -2 & 4 & 0 \\
-6 & 4 & 3 & -8 & 1
\end{bmatrix}
\xrightarrow{\ r_1 \leftrightarrow r_2\ }
\begin{bmatrix}
2 & -2 & 1 & 0 & -1 \\
0 & 1 & -3 & 4 & 1 \\
2 & -1 & -2 & 4 & 0 \\
-6 & 4 & 3 & -8 & 1
\end{bmatrix}
$$

$$
\xrightarrow[\ 3r_1 \text{ added to } r_4\ ]{\ -r_1 \text{ added to } r_3\ }
\begin{bmatrix}
2 & -2 & 1 & 0 & -1 \\
0 & 1 & -3 & 4 & 1 \\
0 & 1 & -3 & 4 & 1 \\
0 & -2 & 6 & -8 & -2
\end{bmatrix}
$$

$$
\xrightarrow[\ 2r_2 \text{ added to } r_4\ ]{\ -r_2 \text{ added to } r_3\ }
\begin{bmatrix}
2 & -2 & 1 & 0 & -1 \\
0 & 1 & -3 & 4 & 1 \\
0 & 0 & 0 & 0 & 0 \\
0 & 0 & 0 & 0 & 0
\end{bmatrix}
$$

Since only 2 nonzero rows remain in this final (echelon) matrix, there are only 2 constraints, and, consequently, $4 - 2 = 2$ of the unknowns—y and z say—are free variables. Let $y = t_1$ and $z = t_2$. Back-substitution of $y = t_1$ and $z = t_2$ into the second row ($x - 3y + 4z = 1$) gives

$$x - 3t_1 + 4t_2 = 1 \quad \Rightarrow \quad x = 1 + 3t_1 - 4t_2$$

Finally, back-substituting $x = 1 + 3t_1 - 4t_2$, $y = t_1$, and $z = t_2$ into the first row ($2w - 2x + y = -1$) determines w:

$$2w - 2(1 + 3t_1 - 4t_2) + t_1 = -1 \quad \Rightarrow \quad w = \tfrac{1}{2} + \tfrac{5}{2}t_1 - 4t_2$$

Therefore, every solution of this system has the form

$$(w, x, y, z) = (\tfrac{1}{2} + \tfrac{5}{2}t_1 - 4t_2, \ 1 + 3t_1 - 4t_2, \ t_1, \ t_2)$$

where t_1 and t_2 are any real numbers. Another way to write the solution is as follows:

$$
\begin{aligned}
(w,\ x,\ y,\ z) &= (\tfrac{1}{2} + \tfrac{5}{2}t_1 - 4t_2,\ 1 + 3t_1 - 4t_2,\ t_1,\ t_2) \\
&= (\tfrac{5}{2}t_1,\ 3t_1,\ t_1,\ 0) + (-4t_2,\ -4t_2,\ 0,\ t_2) + (\tfrac{1}{2},\ 1,\ 0,\ 0) \\
&= t_1(\tfrac{5}{2},\ 3,\ 1,\ 0) + t_2(-4, -4,\ 0,\ 1) + (\tfrac{1}{2},\ 1,\ 0,\ 0)
\end{aligned}
$$

where $t_1,\ t_2 \in \mathbf{R}$. ■

Example 17: Determine the general solution of

$$
\begin{array}{rrrrrrrr}
 & x & - & 3y & + & 4z & = & 0 \\
2w & - & 2x & + & y & & = & 0 \\
2w & - & x & - & 2y & + & 4z & = & 0 \\
-6w & + & 4x & + & 3y & - & 8z & = & 0
\end{array}
$$

which is the homogeneous system corresponding to the non-homogeneous one in Example 16 above.

Since the solution to the nonhomogeneous system in Example 16 is

$$
(w,\ x,\ y,\ z) = \underbrace{t_1(\tfrac{5}{2},\ 3,\ 1,\ 0) + t_2(-4, -4,\ 0,\ 1)}_{\mathbf{x}_h} + \underbrace{(\tfrac{1}{2},\ 1,\ 0,\ 0)}_{\bar{\mathbf{x}}} \quad (*)
$$

Theorem C implies that the solution of the corresponding homogeneous system is

$$
(w,\ x,\ y,\ z) = t_1(\tfrac{5}{2},\ 3,\ 1,\ 0) + t_2(-4, -4,\ 0,\ 1)
$$

(where $t_1,\ t_2 \in \mathbf{R}$), which is obtained from (*) by simply discarding the particular solution, $\bar{\mathbf{x}} = (\tfrac{1}{2},\ 1,\ 0,\ 0)$, of the non-homogeneous system. ■

Example 18: Prove Theorem A: Regardless of its size or the number of unknowns its equations contain, a linear system will have either no solutions, exactly one solution, or infinitely many solutions.

Proof. Let the given linear system be written in matrix form, $A\mathbf{x} = \mathbf{b}$. The theorem really comes down to this: if $A\mathbf{x} = \mathbf{b}$ has more than one solution, then it actually has infinitely many. To establish this, let $\mathbf{x}_1$ and $\mathbf{x}_2$ be two distinct solutions of $A\mathbf{x} = \mathbf{b}$. It will now be shown that for any real value of t, the vector $\mathbf{x}_1 + t(\mathbf{x}_1 - \mathbf{x}_2)$ is also a solution of $A\mathbf{x} = \mathbf{b}$; because t can take on infinitely many different values, the desired conclusion will follow. Since $A\mathbf{x}_1 = \mathbf{b}$ and $A\mathbf{x}_2 = \mathbf{b}$,

$$\begin{aligned} A\big[\mathbf{x}_1 + t(\mathbf{x}_1 - \mathbf{x}_2)\big] &= A\mathbf{x}_1 + A\big[t(\mathbf{x}_1 - \mathbf{x}_2)\big] \\ &= A\mathbf{x}_1 + tA(\mathbf{x}_1 - \mathbf{x}_2) \\ &= A\mathbf{x}_1 + t(A\mathbf{x}_1 - A\mathbf{x}_2) \\ &= \mathbf{b} + t(\mathbf{b} - \mathbf{b}) \\ &= \mathbf{b} \end{aligned}$$

Therefore, $\mathbf{x}_1 + t(\mathbf{x}_1 - \mathbf{x}_2)$ is indeed a solution of $A\mathbf{x} = \mathbf{b}$, and the theorem is proved. ∎

Using Elementary Row Operations to Determine A^{-1}

A linear system is said to be **square** if the number of equations matches the number of unknowns. The systems in Examples 1 and 9 were square, for example. If the system $A\mathbf{x} = \mathbf{b}$ is square, then the coefficient matrix, A, is square. If A has an inverse, then the solution to the system $A\mathbf{x} = \mathbf{b}$ can be found by multiplying both sides by A^{-1}:

$$A\mathbf{x} = \mathbf{b} \quad \Rightarrow \quad A^{-1}A\mathbf{x} = A^{-1}\mathbf{b} \quad \Rightarrow \quad \mathbf{x} = A^{-1}\mathbf{b}$$

This calculation establishes the following result:

Theorem D. If A is an invertible n by n matrix, then the system $A\mathbf{x} = \mathbf{b}$ has a unique solution for *every* n-vector $\mathbf{b}$, and this solution equals $A^{-1}\mathbf{b}$.

Since the determination of A^{-1} typically requires more calculation than performing Gaussian elimination and back-substitution, this is not necessarily an improved method of solving $A\mathbf{x} = \mathbf{b}$. (And, of course, if A is not square, then it has no inverse, so this method is not even an option for nonsquare systems.) However, if the coefficient matrix A is square, and if A^{-1} is known or the solution of $A\mathbf{x} = \mathbf{b}$ is required for several different $\mathbf{b}$'s, then this method is indeed useful, from both a theoretical and a practical point of view. The purpose of this section is to show how the elementary row operations that characterize Gauss-Jordan elimination can be applied to compute the inverse of a square matrix.

First, a definition: If an elementary row operation (the interchange of two rows, the multiplication of a row by a nonzero constant, or the addition of a multiple of one row to another) is applied to the identity matrix, I, the result is called an **elementary matrix**. To illustrate, consider the 3 by 3 identity matrix. If the first and third rows are interchanged,

$$\begin{bmatrix} 1 & & \\ & 1 & \\ & & 1 \end{bmatrix} \xrightarrow{\ \mathbf{r}_1 \leftrightarrow \mathbf{r}_3\ } \begin{bmatrix} & & 1 \\ & 1 & \\ 1 & & \end{bmatrix}$$

or if the second row of I is multiplied by -2,

$$\begin{bmatrix} 1 & & \\ & 1 & \\ & & 1 \end{bmatrix} \xrightarrow{-2\mathbf{r}_2} \begin{bmatrix} 1 & & \\ & -2 & \\ & & 1 \end{bmatrix}$$

or if -2 times the first row is added to the second row,

$$\begin{bmatrix} 1 & & \\ & 1 & \\ & & 1 \end{bmatrix} \xrightarrow{-2\mathbf{r}_1 \text{ added to } \mathbf{r}_2} \begin{bmatrix} 1 & & \\ -2 & 1 & \\ & & 1 \end{bmatrix}$$

all of these resulting matrices are examples of elementary matrices. The first fact that will be needed to compute A^{-1} reads as follows: *If E is the elementary matrix that results when a particular elementary row operation is performed on I, then the product EA is equal to the matrix that would result if that same elementary row operation were applied to A.* In other words, an elementary row operation on a matrix A can be performed by multiplying A on the left by the corresponding elementary matrix. For example, consider the matrix

$$A = \begin{bmatrix} 1 & -1 & 2 \\ 2 & 0 & 3 \\ 0 & 1 & -1 \end{bmatrix}$$

Adding -2 times the first row to the second row yields

$$A = \begin{bmatrix} 1 & -1 & 2 \\ 2 & 0 & 3 \\ 0 & 1 & -1 \end{bmatrix} \xrightarrow{-2\mathbf{r}_1 \text{ added to } \mathbf{r}_2} \begin{bmatrix} 1 & -1 & 2 \\ 0 & 2 & -1 \\ 0 & 1 & -1 \end{bmatrix} = A'$$

If this same elementary row operation is applied to I,

$$I = \begin{bmatrix} 1 & & \\ & 1 & \\ & & 1 \end{bmatrix} \xrightarrow{\ -2\mathbf{r}_1 \text{ added to } \mathbf{r}_2\ } \begin{bmatrix} 1 & & \\ -2 & 1 & \\ & & 1 \end{bmatrix} = E$$

then the result above guarantees that EA should equal A'. You may verify that

$$EA = \begin{bmatrix} 1 & & \\ -2 & 1 & \\ & & 1 \end{bmatrix} \begin{bmatrix} 1 & -1 & 2 \\ 2 & 0 & 3 \\ 0 & 1 & -1 \end{bmatrix} = \begin{bmatrix} 1 & -1 & 2 \\ 0 & 2 & -1 \\ 0 & 1 & -1 \end{bmatrix} = A'$$

is indeed true.

If A is an invertible matrix, then some sequence of elementary row operations will transform A into the identity matrix, I. Since each of these operations is equivalent to left multiplication by an elementary matrix, the first step in the reduction of A to I would be given by the product E_1A, the second step would be given by E_2E_1A, and so on. Thus, there exist elementary matrices $E_1, E_2, \ldots, E_k$ such that

$$E_k \cdots E_2 E_1 A = I$$

But this equation makes it clear that $E_k \cdots E_2 E_1 = A^{-1}$:

$$\underbrace{E_k \cdots E_2 E_1}_{A^{-1}} A = I$$

Since $E_k \cdots E_2 E_1 = E_k \cdots E_2 E_1 I$, where the right-hand side explicitly denotes the elementary row operations applied to the identity matrix I, *the same elementary row operations that transform A into I will transform I into A^{-1}.* For n by n matrices A with $n \geq 3$, this describes the most efficient method for determining A^{-1}.

Example 19: Determine the inverse of the matrix

$$A = \begin{bmatrix} 1 & -1 & 2 \\ 2 & 0 & 3 \\ 0 & 1 & -1 \end{bmatrix}$$

Since the elementary row operations that will be applied to A will be applied to I as well, it is convenient here to augment the matrix A with the identity matrix I:

$$[A|I] = \begin{bmatrix} 1 & -1 & 2 & | & 1 & 0 & 0 \\ 2 & 0 & 3 & | & 0 & 1 & 0 \\ 0 & 1 & -1 & | & 0 & 0 & 1 \end{bmatrix}$$

Then, as A is transformed into I, I will be transformed into A^{-1}:

$$[A|I] \longrightarrow [I|A^{-1}]$$

Now for a sequence of elementary row operations that will effect this transformation:

$$\begin{bmatrix} 1 & -1 & 2 & 1 & 0 & 0 \\ 2 & 0 & 3 & 0 & 1 & 0 \\ 0 & 1 & -1 & 0 & 0 & 1 \end{bmatrix}$$

$$\xrightarrow{-2\mathbf{r}_1 \text{ added to } \mathbf{r}_2} \begin{bmatrix} 1 & -1 & 2 & 1 & 0 & 0 \\ 0 & 2 & -1 & -2 & 1 & 0 \\ 0 & 1 & -1 & 0 & 0 & 1 \end{bmatrix}$$

$$\xrightarrow{\mathbf{r}_2 \leftrightarrow \mathbf{r}_3} \begin{bmatrix} 1 & -1 & 2 & 1 & 0 & 0 \\ 0 & 1 & -1 & 0 & 0 & 1 \\ 0 & 2 & -1 & -2 & 1 & 0 \end{bmatrix}$$

$$\xrightarrow{-2\mathbf{r}_2 \text{ added to } \mathbf{r}_3} \begin{bmatrix} 1 & -1 & 2 & 1 & 0 & 0 \\ 0 & 1 & -1 & 0 & 0 & 1 \\ 0 & 0 & 1 & -2 & 1 & -2 \end{bmatrix}$$

$$\xrightarrow[\begin{subarray}{c} \mathbf{r}_3 \text{ added to } \mathbf{r}_2 \\ -2\mathbf{r}_3 \text{ added to } \mathbf{r}_1 \end{subarray}]{} \begin{bmatrix} 1 & -1 & 0 & 5 & -2 & 4 \\ 0 & 1 & 0 & -2 & 1 & -1 \\ 0 & 0 & 1 & -2 & 1 & -2 \end{bmatrix}$$

$$\xrightarrow{\mathbf{r}_2 \text{ added to } \mathbf{r}_1} \begin{bmatrix} 1 & 0 & 0 & 3 & -1 & 3 \\ 0 & 1 & 0 & -2 & 1 & -1 \\ 0 & 0 & 1 & -2 & 1 & -2 \end{bmatrix}$$

Since the transformation $[A \mid I] \rightarrow [I \mid A^{-1}]$ reads

$$\underbrace{\begin{bmatrix} 1 & -1 & 2 \\ 2 & 0 & 3 \\ 0 & 1 & -1 \end{bmatrix}}_{A} \underbrace{\begin{bmatrix} 1 & 0 & 0 \\ 0 & 1 & 0 \\ 0 & 0 & 1 \end{bmatrix}}_{I} \longrightarrow \underbrace{\begin{bmatrix} 1 & 0 & 0 \\ 0 & 1 & 0 \\ 0 & 0 & 1 \end{bmatrix}}_{I} \underbrace{\begin{bmatrix} 3 & -1 & 3 \\ -2 & 1 & -1 \\ -2 & 1 & -2 \end{bmatrix}}_{A^{-1}}$$

the inverse of the given matrix A is

$$A^{-1} = \begin{bmatrix} 3 & -1 & 3 \\ -2 & 1 & -1 \\ -2 & 1 & -2 \end{bmatrix} \quad \blacksquare$$

Example 20: What condition must the entries of a general 2 by 2 matrix

$$A = \begin{bmatrix} a & b \\ c & d \end{bmatrix}$$

satisfy in order for A to be invertible? What is the inverse of A in this case?

The goal is to effect the transformation $[A \mid I] \rightarrow [I \mid A^{-1}]$. First, augment A with the 2 by 2 identity matrix:

$$[A \quad I] = \begin{bmatrix} a & b & 1 & 0 \\ c & d & 0 & 1 \end{bmatrix}$$

Now, if $a = 0$, switch the rows. If c is also 0, then the process of reducing A to I cannot even begin. So, one necessary condition for A to be invertible is that the entries a and c are not both 0. Assume that $a \neq 0$. Then

$$\begin{bmatrix} a & b & 1 & 0 \\ c & d & 0 & 1 \end{bmatrix} \xrightarrow{\frac{1}{a}\mathbf{r}_1} \begin{bmatrix} 1 & \frac{b}{a} & \frac{1}{a} & 0 \\ c & d & 0 & 1 \end{bmatrix}$$

$$\xrightarrow{-c\mathbf{r}_1 \text{ added to } \mathbf{r}_2} \begin{bmatrix} 1 & \frac{b}{a} & \frac{1}{a} & 0 \\ 0 & \frac{ad-bc}{a} & -\frac{c}{a} & 1 \end{bmatrix}$$

Next, *assuming that* $ad - bc \neq 0$,

$$\begin{bmatrix} 1 & \frac{b}{a} & \frac{1}{a} & 0 \\ 0 & \frac{ad-bc}{a} & -\frac{c}{a} & 1 \end{bmatrix}$$

$$\xrightarrow{\frac{-b}{ad-bc}\mathbf{r}_2 \text{ added to } \mathbf{r}_1} \begin{bmatrix} 1 & 0 & \frac{d}{ad-bc} & \frac{-b}{ad-bc} \\ 0 & \frac{ad-bc}{a} & -\frac{c}{a} & 1 \end{bmatrix}$$

$$\xrightarrow{\frac{a}{ad-bc}\mathbf{r}_2} \begin{bmatrix} 1 & 0 & \frac{d}{ad-bc} & \frac{-b}{ad-bc} \\ 0 & 1 & -\frac{c}{ad-bc} & \frac{a}{ad-bc} \end{bmatrix}$$

Therefore, if $ad - bc \neq 0$, then the matrix A is invertible, and its inverse is given by

$$\begin{bmatrix} a & b \\ c & d \end{bmatrix}^{-1} = \frac{1}{ad-bc} \begin{bmatrix} d & -b \\ -c & a \end{bmatrix}$$

(The requirement that a and c are not both 0 is automatically included in the condition $ad - bc \neq 0$.) In words, the inverse is obtained from the given matrix by interchanging the diagonal entries, changing the signs of the off-diagonal entries, and then dividing by the quantity $ad - bc$. *This formula for the inverse of a 2 × 2 matrix should be memorized.*

To illustrate, consider the matrix

$$A = \begin{bmatrix} -2 & -3 \\ 4 & 5 \end{bmatrix}$$

Since $ad - bc = (-2)(5) - (-3)(4) = 2 \neq 0$, the matrix is invertible, and its inverse is

$$A^{-1} = \tfrac{1}{2}\begin{bmatrix} 5 & 3 \\ -4 & -2 \end{bmatrix} = \begin{bmatrix} \tfrac{5}{2} & \tfrac{3}{2} \\ -2 & -1 \end{bmatrix}$$

You may verify that

$$AA^{-1} = \begin{bmatrix} -2 & -3 \\ 4 & 5 \end{bmatrix}\begin{bmatrix} \tfrac{5}{2} & \tfrac{3}{2} \\ -2 & -1 \end{bmatrix} = \begin{bmatrix} 1 & 0 \\ 0 & 1 \end{bmatrix} = I$$

and that $A^{-1}A = I$ also. ∎

Example 21: Let A be the matrix

$$\begin{bmatrix} 2 & 1 & -1 \\ -1 & -3 & 1 \\ 1 & 8 & -2 \end{bmatrix}$$

in Example 14 above. Is A invertible?

No. Recall that row reduction of A produced the matrix

$$A' = \begin{bmatrix} -1 & -3 & 1 \\ 0 & -5 & 1 \\ 0 & 0 & 0 \end{bmatrix}$$

The row of zeros signifies that A cannot be transformed to the identity matrix by a sequence of elementary row operations; A is noninvertible. Another argument for the noninvertibility of A follows from the result of Example 14 and Theorem D. If A were invertible, then Theorem D would guarantee the existence of a solution to $A\mathbf{x} = \mathbf{b}$ for *every* column vector $\mathbf{b} = (b_1, b_2, b_3)^T$. But Example 14 showed that $A\mathbf{x} = \mathbf{b}$ is consistent only for those vectors $\mathbf{b}$ for which $b_1 + 3b_2 + b_3 = 0$. Clearly, then, there exist (infinitely many) vectors $\mathbf{b}$ for which $A\mathbf{x} = \mathbf{b}$ is inconsistent; thus, A cannot be invertible. ∎

Example 22: What can you say about the solutions of the homogeneous system $A\mathbf{x} = \mathbf{0}$ if the matrix A is invertible?

Theorem D guarantees that for an invertible matrix A, the system $A\mathbf{x} = \mathbf{b}$ is consistent for every possible choice of the column vector $\mathbf{b}$ and that the unique solution is given by $A^{-1}\mathbf{b}$. In the case of a homogeneous system, the vector $\mathbf{b}$ is $\mathbf{0}$, so the system has only the trivial solution: $\mathbf{x} = A^{-1}\mathbf{0} = \mathbf{0}$. ∎

Example 23: Solve the matrix equation $AX = B$, where

$$A = \begin{bmatrix} 1 & 4 & -2 \\ -1 & 1 & -1 \\ 3 & 0 & 1 \end{bmatrix} \quad \text{and} \quad B = \begin{bmatrix} 1 & 12 \\ -7 & 2 \\ 17 & 3 \end{bmatrix}$$

Solution 1. Since A is 3×3 and B is 3×2, if a matrix X exists such that $AX = B$, then X must be 3×2. If A is invertible, one way to find X is to determine A^{-1} and then to compute $X = A^{-1}B$. The algorithm $[A \mid I] \rightarrow [I \mid A^{-1}]$ to find A^{-1} yields

$$\begin{bmatrix} 1 & 4 & -2 & 1 & 0 & 0 \\ -1 & 1 & -1 & 0 & 1 & 0 \\ 3 & 0 & 1 & 0 & 0 & 1 \end{bmatrix}$$

$$\xrightarrow[\substack{r_1 \text{ added to } r_2 \\ -3r_1 \text{ added to } r_3}]{} \begin{bmatrix} 1 & 4 & -2 & 1 & 0 & 0 \\ 0 & 5 & -3 & 1 & 1 & 0 \\ 0 & -12 & 7 & -3 & 0 & 1 \end{bmatrix}$$

$$\xrightarrow[\substack{2r_2 \text{ added to } r_3}]{} \begin{bmatrix} 1 & 4 & -2 & 1 & 0 & 0 \\ 0 & 5 & -3 & 1 & 1 & 0 \\ 0 & -2 & 1 & -1 & 2 & 1 \end{bmatrix}$$

$$\xrightarrow[\substack{2r_3 \text{ added to } r_2}]{} \begin{bmatrix} 1 & 4 & -2 & 1 & 0 & 0 \\ 0 & 1 & -1 & -1 & 5 & 2 \\ 0 & -2 & 1 & -1 & 2 & 1 \end{bmatrix}$$

$$\xrightarrow[\substack{2r_2 \text{ added to } r_3}]{} \begin{bmatrix} 1 & 4 & -2 & 1 & 0 & 0 \\ 0 & 1 & -1 & -1 & 5 & 2 \\ 0 & 0 & -1 & -3 & 12 & 5 \end{bmatrix}$$

$$\xrightarrow[\substack{-r_3 \text{ added to } r_2 \\ -2r_3 \text{ added to } r_1}]{} \begin{bmatrix} 1 & 4 & 0 & 7 & -24 & -10 \\ 0 & 1 & 0 & 2 & -7 & -3 \\ 0 & 0 & -1 & -3 & 12 & 5 \end{bmatrix}$$

$$\xrightarrow[\substack{-4r_2 \text{ added to } r_1 \\ -r_3}]{} \begin{bmatrix} 1 & 0 & 0 & -1 & 4 & 2 \\ 0 & 1 & 0 & 2 & -7 & -3 \\ 0 & 0 & 1 & 3 & -12 & -5 \end{bmatrix}$$

Therefore,

$$A^{-1} = \begin{bmatrix} -1 & 4 & 2 \\ 2 & -7 & -3 \\ 3 & -12 & -5 \end{bmatrix}$$

so

$$X = A^{-1}B = \begin{bmatrix} -1 & 4 & 2 \\ 2 & -7 & -3 \\ 3 & -12 & -5 \end{bmatrix} \begin{bmatrix} 1 & 12 \\ -7 & 2 \\ 17 & 3 \end{bmatrix} = \begin{bmatrix} 5 & 2 \\ 0 & 1 \\ 2 & -3 \end{bmatrix}$$

Solution 2. Let $\mathbf{b}_1$ and $\mathbf{b}_2$ denote, respectively, column 1 and column 2 of the matrix B. If the solution to $A\mathbf{x} = \mathbf{b}_1$ is $\mathbf{x}_1$ and the solution to $A\mathbf{x} = \mathbf{b}_2$ is $\mathbf{x}_2$, then the solution to $AX = B = [\mathbf{b}_1 \ \mathbf{b}_2]$ is $X = [\mathbf{x}_1 \ \mathbf{x}_2]$. That is, the elimination procedure can be performed on the two systems ($A\mathbf{x} = \mathbf{b}_1$ and $A\mathbf{x} = \mathbf{b}_2$) simultaneously:

$$\begin{bmatrix} 1 & 4 & -2 & 1 & 12 \\ -1 & 1 & -1 & -7 & 2 \\ 3 & 0 & 1 & 17 & 3 \end{bmatrix}$$

$$\xrightarrow[\substack{\mathbf{r}_1 \text{ added to } \mathbf{r}_2 \\ -3\mathbf{r}_1 \text{ added to } \mathbf{r}_3}]{} \begin{bmatrix} 1 & 4 & -2 & 1 & 12 \\ 0 & 5 & -3 & -6 & 14 \\ 0 & -12 & 7 & 14 & -33 \end{bmatrix}$$

$$\xrightarrow[2\mathbf{r}_2 \text{ added to } \mathbf{r}_3]{} \begin{bmatrix} 1 & 4 & -2 & 1 & 12 \\ 0 & 5 & -3 & -6 & 14 \\ 0 & -2 & 1 & 2 & -5 \end{bmatrix}$$

$$\xrightarrow{\text{2r}_3 \text{ added to r}_2} \begin{bmatrix} 1 & 4 & -2 & 1 & 12 \\ 0 & 1 & -1 & -2 & 4 \\ 0 & -2 & 1 & 2 & -5 \end{bmatrix}$$

$$\xrightarrow{\text{2r}_2 \text{ added to r}_3} \begin{bmatrix} 1 & 4 & -2 & 1 & 12 \\ 0 & 1 & -1 & -2 & 4 \\ 0 & 0 & -1 & -2 & 3 \end{bmatrix}$$

Gauss-Jordan elimination completes the evaluation of the components of x_1 and x_2:

$$\begin{bmatrix} 1 & 4 & -2 & 1 & 12 \\ 0 & 1 & -1 & -2 & 4 \\ 0 & 0 & -1 & -2 & 3 \end{bmatrix}$$

$$\xrightarrow[\substack{-2\text{r}_3 \text{ added to r}_1}]{-\text{r}_3 \text{ added to r}_2} \begin{bmatrix} 1 & 4 & 0 & 5 & 6 \\ 0 & 1 & 0 & 0 & 1 \\ 0 & 0 & -1 & -2 & 3 \end{bmatrix}$$

$$\xrightarrow[\substack{-\text{r}_3}]{-4\text{r}_2 \text{ added to r}_1} \begin{bmatrix} 1 & 0 & 0 & 5 & 2 \\ 0 & 1 & 0 & 0 & 1 \\ 0 & 0 & 1 & 2 & -3 \end{bmatrix}$$

It follows immediately from this final augmented matrix that

$$X = \begin{bmatrix} 5 & 2 \\ 0 & 1 \\ 2 & -3 \end{bmatrix}$$

as before.

It is easy to verify that the matrix X does indeed satisfy the equation $AX = B$:

$$AX = \begin{bmatrix} 1 & 4 & -2 \\ -1 & 1 & -1 \\ 3 & 0 & 1 \end{bmatrix} \begin{bmatrix} 5 & 2 \\ 0 & 1 \\ 2 & -3 \end{bmatrix} = \begin{bmatrix} 1 & 12 \\ -7 & 2 \\ 17 & 3 \end{bmatrix} = B$$

Note that the transformation in Solution 1 was $[A \mid I] \rightarrow [I \mid A^{-1}]$, from which $A^{-1}B$ was computed to give X. However, the transformation in Solution 2, $[A \mid B] \rightarrow [I \mid X]$, gave X directly. ∎

The concept of a vector space is of fundamental importance throughout much of mathematics and physics. Although the most general definition of a vector space is not needed in an introduction to linear algebra, the particular type of vector space that will be studied here—the Euclidean vector space—is the one most frequently used in applications of the subject. Since all scalars in this book are real, the resulting structure is called a *real* Euclidean vector space.

Subspaces of $\mathbf{R}^n$

Consider the collection of vectors

$$V = \left\{ (x,\ 3x) \colon x \in \mathbf{R} \right\}$$

The endpoints of all such vectors lie on the line $y = 3x$ in the x-y plane. Now, choose any two vectors from V, say, $\mathbf{u} = (1,\ 3)$ and $\mathbf{v} = (-2,\ -6)$. Note that the sum of $\mathbf{u}$ and $\mathbf{v}$,

$$\mathbf{u} + \mathbf{v} = (-1,\ -3)$$

is also a vector in V, because its second component is three times the first. In fact, it can be easily shown that the sum of *any* two vectors in V will produce a vector that again lies in V. The set V is therefore said to be **closed under addition**. Next, consider a scalar multiple of $\mathbf{u}$, say,

$$5\mathbf{u} = 5(1,\ 3) = (5,\ 15)$$

It, too, is in V. In fact, *every* scalar multiple of any vector in V is itself an element of V. The set V is therefore said to be **closed under scalar multiplication**.

Thus, the elements in V enjoy the following two properties:

(1) Closure under addition:

 The sum of any two elements in V is an element of V.

(2) Closure under scalar multiplication:

 Every scalar multiple of an element in V is an element of V.

Any subset of $\mathbf{R}^n$ that satisfies these two properties—with the usual operations of addition and scalar multiplication—is called a **subspace of $\mathbf{R}^n$** or a **Euclidean vector space**. The set $V = \{(x, 3x): x \in \mathbf{R}\}$ is a Euclidean vector space, a subspace of $\mathbf{R}^2$.

Example 1: Is the following set a subspace of $\mathbf{R}^2$?

$$A = \{(x, 3x+1): x \in \mathbf{R}\}$$

To establish that A is a subspace of $\mathbf{R}^2$, it must be shown that A is closed under addition and scalar multiplication. If a counterexample to even one of these properties can be found, then the set is not a subspace. In the present case, it is very easy to find such a counterexample. For instance, both $\mathbf{u} = (1, 4)$ and $\mathbf{v} = (2, 7)$ are in A, but their sum, $\mathbf{u} + \mathbf{v} = (3, 11)$, is not. In order for a vector $\mathbf{v} = (v_1, v_2)$ to be in A, the second component (v_2) must be 1 more than three times the first component (v_1). Since $11 \neq 3(3) + 1$, $(3, 11) \notin A$. Therefore, the set A is not closed under addition, so A cannot be a subspace. [You could also show that this particular set is not a subspace of $\mathbf{R}^2$ by exhibiting a counterexample to closure under scalar multiplication. For example, although $\mathbf{u} = (1, 4)$ is in A, the scalar multiple $2\mathbf{u} = (2, 8)$ is not.] ■

Example 2: Is the following set a subspace of $\mathbf{R}^3$?

$$B = \left\{ (x, x^2, x^3): x \in \mathbf{R} \right\}$$

In order for a subset of $\mathbf{R}^3$ to be a subspace of $\mathbf{R}^3$, both closure properties (1) and (2) must be satisfied. However, note that while $\mathbf{u} = (1, 1, 1)$ and $\mathbf{v} = (2, 4, 8)$ are both in B, their sum, $(3, 5, 9)$, clearly is not. Since B is not closed under addition, B is not a subspace of $\mathbf{R}^3$. ■

Example 3: Is the following set a subspace of $\mathbf{R}^4$?

$$C = \left\{ (x_1, 0, x_3, -5x_1): x_1, x_3 \in \mathbf{R} \right\}$$

For a 4-vector to be in C, exactly two conditions must be satisfied: Namely, its second component must be zero, and its fourth component must be −5 times the first. Choosing particular vectors in C and checking closure under addition and scalar multiplication would lead you to conjecture that C is indeed a subspace. However, no matter how many specific examples you provide showing that the closure properties are satisfied, the fact that C is a subspace is established only when a general proof is given. So let $\mathbf{u} = (u_1, 0, u_3, -5u_1)$ and $\mathbf{v} = (v_1, 0, v_3, -5v_1)$ be arbitrary vectors in C. Then their sum,

$$\mathbf{u} + \mathbf{v} = \left(u_1 + v_1, 0, u_3 + v_3, -5(u_1 + v_1) \right)$$

satisfies the conditions for membership in C, verifying closure under addition. Finally, if k is a scalar, then

$$k\mathbf{u} = \left(ku_1, 0, ku_3, -5(ku_1) \right)$$

is in C, establishing closure under scalar multiplication. This proves that C is a subspace of $\mathbf{R}^4$. ∎

Example 4: Show that if V is a subspace of $\mathbf{R}^n$, then V must contain the zero vector.

First, choose any vector $\mathbf{v}$ in V. Since V is a subspace, it must be closed under scalar multiplication. By selecting 0 as the scalar, the vector $0\mathbf{v}$, which equals $\mathbf{0}$, must be in V. [Another method proceeds like this: If $\mathbf{v}$ is in V, then the scalar multiple $(-1)\mathbf{v} = -\mathbf{v}$ must also be in V. But then the sum of these two vectors, $\mathbf{v} + (-\mathbf{v}) = \mathbf{0}$, must be in V, since V is closed under addition.]

This result can provide a quick way to conclude that a particular set is not a Euclidean space. *If the set does not contain the zero vector, then it cannot be a subspace.* For example, the set A in Example 1 above could not be a subspace of $\mathbf{R}^2$ because it does not contain the vector $\mathbf{0} = (0, 0)$. It is important to realize that containing the zero vector is a *necessary* condition for a set to be a Euclidean space, not a *sufficient* one. That is, just because a set contains the zero vector does not guarantee that it is a Euclidean space (for example, consider the set B in Example 2); the guarantee is that if the set does *not* contain $\mathbf{0}$, then it is *not* a Euclidean vector space. ∎

As always, the distinction between vectors and points can be blurred, and sets consisting of points in $\mathbf{R}^n$ can be considered for classification as subspaces.

Example 5: Is the following set a subspace of $\mathbf{R}^2$?

$$D = \{(x, y): x \geq 0 \text{ and } y \geq 0\}$$

As illustrated in Figure 40, this set consists of all points in the first quadrant, including the points $(x, 0)$ on the x axis with $x \geq 0$ and the points $(0, y)$ on the y axis with $y \geq 0$:

■   Figure 40   ■

The set D is closed under addition since the sum of nonnegative numbers is nonnegative. That is, if (x_1, y_1) and (x_2, y_2) are in D, then x_1, x_2, y_1, and y_2 are all greater than or equal to 0, so both sums $x_1 + x_2$ and $y_1 + y_2$ are greater than or equal to 0. This implies that

$$(x_1, y_1) + (x_2, y_2) = (x_1 + x_2, y_1 + y_2) \in D$$

However, D is not closed under scalar multiplication. If x and y are both positive, then (x, y) is in D, but for any negative scalar k,

$$k(x, y) = (kx, ky) \notin D$$

since $kx < 0$ (and $ky < 0$). Therefore, D is not a subspace of $\mathbf{R}^2$. ■

Example 6: Is the following set a subspace of $\mathbf{R}^2$?

$$E = \{(x, y): xy \geq 0\}$$

As illustrated in Figure 41, this set consists of all points in the first and third quadrants, including the axes:

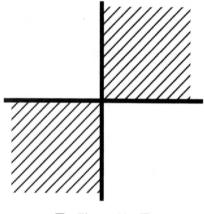

■ Figure 41 ■

The set E is closed under scalar multiplication, since if k is any scalar, then $k(x, y) = (kx, ky)$ is in E. The proof of this last statement follows immediately from the condition for membership in E. A point is in E if the product of its two coordinates is nonnegative. Since $k^2 \geq 0$ for any real k,

$$(x, y) \in E \;\Rightarrow\; xy \geq 0 \;\Rightarrow\; k^2 xy = (kx)(ky) \geq 0$$
$$\Rightarrow\; (kx, ky) \in E$$

However, although E is closed under scalar multiplication, it is not closed under addition. For example, although $\mathbf{u} = (4, 1)$ and $\mathbf{v} = (-2, -6)$ are both in E, their sum, $(2, -5)$, is not. Thus, E is not a subspace of $\mathbf{R}^2$. ∎

Example 7: Does the plane P given by the equation $2x + y - 3z = 0$ form a subspace of $\mathbf{R}^3$?

One way to characterize P is to solve the given equation for y,

$$y = 3z - 2x$$

and write

$$P = \{(x, 3z - 2x, z): x, z \in \mathbf{R}\}$$

If $\mathbf{p}_1 = (x_1, 3z_1 - 2x_1, z_1)$ and $\mathbf{p}_2 = (x_2, 3z_2 - 2x_2, z_2)$ are points in P, then their sum,

$$\mathbf{p}_1 + \mathbf{p}_2 = \left(x_1 + x_2, \ 3(z_1 + z_2) - 2(x_1 + x_2), \ z_1 + z_2\right)$$

is also in P, so P is closed under addition. Furthermore, if $\mathbf{p} = (x, 3z - 2x, z)$ is a point in P, then any scalar multiple,

$$k\mathbf{p} = \left(kx, \ 3(kz) - 2(kx), \ kz\right)$$

is also in P, so P is also closed under scalar multiplication. Therefore, P does indeed form a subspace of $\mathbf{R}^3$. Note that P contains the origin. By contrast, the plane $2x + y - 3z = 1$, although parallel to P, is *not* a subspace of $\mathbf{R}^3$ because it does not contain $(0, 0, 0)$; recall Example 4 above. In fact, a plane in $\mathbf{R}^3$ is a subspace of $\mathbf{R}^3$ if and only if it contains the origin. ∎

The Nullspace of a Matrix

The solution sets of homogeneous linear systems provide an important source of vector spaces. Let A be an m by n matrix, and consider the homogeneous system

$$A\mathbf{x} = \mathbf{0}$$

Since A is m by n, the set of all vectors $\mathbf{x}$ which satisfy this equation forms a subset of $\mathbf{R}^n$. (This subset is nonempty, since it clearly contains the zero vector: $\mathbf{x} = \mathbf{0}$ always satisfies $A\mathbf{x} = \mathbf{0}$.) This subset actually forms a subspace of $\mathbf{R}^n$, called the **nullspace** of the matrix A and denoted $N(A)$. To prove that $N(A)$ is a subspace of $\mathbf{R}^n$, closure under both addition and scalar multiplication must be established. If $\mathbf{x}_1$ and $\mathbf{x}_2$ are in $N(A)$, then, by definition, $A\mathbf{x}_1 = \mathbf{0}$ and $A\mathbf{x}_2 = \mathbf{0}$. Adding these equations yields

$$A\mathbf{x}_1 + A\mathbf{x}_2 = \mathbf{0} \implies A(\mathbf{x}_1 + \mathbf{x}_2) = \mathbf{0} \implies \mathbf{x}_1 + \mathbf{x}_2 \in N(A)$$

which verifies closure under addition. Next, if $\mathbf{x}$ is in $N(A)$, then $A\mathbf{x} = \mathbf{0}$, so if k is any scalar,

$$k(A\mathbf{x}) = \mathbf{0} \implies A(k\mathbf{x}) = \mathbf{0} \implies k\mathbf{x} \in N(A)$$

verifying closure under scalar multiplication. Thus, the solution set of a homogeneous linear system forms a vector space. Note carefully that if the system is *not* homogeneous, then the set of solutions is *not* a vector space since the set will not contain the zero vector.

Example 8: The plane P in Example 7, given by $2x + y - 3z = 0$, was shown to be a subspace of $\mathbf{R}^3$. Another proof that this defines a subspace of $\mathbf{R}^3$ follows from the observation that $2x + y - 3z = 0$ is equivalent to the homogeneous system

$$A \begin{bmatrix} x \\ y \\ z \end{bmatrix} = 0$$

where A is the 1×3 matrix $[2 \quad 1 \quad -3]$. P is the nullspace of A. ∎

Example 9: The set of solutions of the homogeneous system

$$\begin{bmatrix} -1 & 1 & 2 & 4 \\ 2 & 0 & 1 & -7 \end{bmatrix} \mathbf{x} = \begin{bmatrix} 0 \\ 0 \end{bmatrix}$$

forms a subspace of $\mathbf{R}^n$ for some n. State the value of n and explicitly determine this subspace.

Since the coefficient matrix is 2 by 4, $\mathbf{x}$ must be a 4-vector. Thus, $n = 4$: The nullspace of this matrix is a subspace of $\mathbf{R}^4$. To determine this subspace, the equation is solved by first row-reducing the given matrix:

$$\begin{bmatrix} -1 & 1 & 2 & 4 \\ 2 & 0 & 1 & -7 \end{bmatrix} \xrightarrow[\ (-1)\mathbf{r}_1\ ]{2\mathbf{r}_1 \text{ added to } \mathbf{r}_2} \begin{bmatrix} 1 & -1 & -2 & -4 \\ 0 & 2 & 5 & 1 \end{bmatrix}$$

Therefore, the system is equivalent to

$$\begin{bmatrix} 1 & -1 & -2 & -4 \\ 0 & 2 & 5 & 1 \end{bmatrix} \begin{bmatrix} x_1 \\ x_2 \\ x_3 \\ x_4 \end{bmatrix} = \begin{bmatrix} 0 \\ 0 \end{bmatrix}$$

that is,

$$x_1 - x_2 - 2x_3 - 4x_4 = 0$$
$$2x_2 + 5x_3 + x_4 = 0$$

If you let x_3 and x_4 be free variables, the second equation directly above implies

$$x_2 = -\tfrac{1}{2}(5x_3 + x_4)$$

Substituting this result into the other equation determines x_1:

$$x_1 - [-\tfrac{1}{2}(5x_3 + x_4)] - 2x_3 - 4x_4 = 0$$
$$x_1 = -\tfrac{1}{2}(x_3 - 7x_4)$$

Therefore, the set of solutions of the given homogeneous system can be written as

$$\left\{ \begin{bmatrix} -\tfrac{1}{2}(x_3 - 7x_4) \\ -\tfrac{1}{2}(5x_3 + x_4) \\ x_3 \\ x_4 \end{bmatrix} : x_3,\ x_4 \in \mathbf{R} \right\}$$

which is a subspace of $\mathbf{R}^4$. This is the nullspace of the matrix

$$\begin{bmatrix} -1 & 1 & 2 & 4 \\ 2 & 0 & 1 & -7 \end{bmatrix} \quad ■$$

Example 10: Find the nullspace of the matrix

$$A = \begin{bmatrix} 2 & 1 \\ 1 & 2 \end{bmatrix}$$

By definition, the nullspace of A consists of all vectors $\mathbf{x}$ such that $A\mathbf{x} = \mathbf{0}$. Perform the following elementary row operations on A,

$$\begin{bmatrix} 2 & 1 \\ 1 & 2 \end{bmatrix} \xrightarrow{\ \mathbf{r}_1 \leftrightarrow \mathbf{r}_2\ } \begin{bmatrix} 1 & 2 \\ 2 & 1 \end{bmatrix} \xrightarrow{\ -2\mathbf{r}_1 \text{ added to } \mathbf{r}_2\ } \begin{bmatrix} 1 & 2 \\ 0 & -3 \end{bmatrix}$$

to conclude that $A\mathbf{x} = \mathbf{0}$ is equivalent to the simpler system

$$\begin{bmatrix} 1 & 2 \\ 0 & -3 \end{bmatrix}\begin{bmatrix} x_1 \\ x_2 \end{bmatrix} = \begin{bmatrix} 0 \\ 0 \end{bmatrix}$$

The second row implies that $x_2 = 0$, and back-substituting this into the first row implies that $x_1 = 0$ also. Since the only solution of $A\mathbf{x} = \mathbf{0}$ is $\mathbf{x} = \mathbf{0}$, the nullspace of A consists of the zero vector alone. This subspace, $\{\mathbf{0}\}$, is called the **trivial subspace** (of $\mathbf{R}^2$). ■

Example 11: Find the nullspace of the matrix

$$B = \begin{bmatrix} 2 & 1 \\ -4 & -2 \end{bmatrix}$$

To solve $B\mathbf{x} = \mathbf{0}$, begin by row-reducing B:

$$\begin{bmatrix} 2 & 1 \\ -4 & -2 \end{bmatrix} \xrightarrow{\ 2\mathbf{r}_1 \text{ added to } \mathbf{r}_2\ } \begin{bmatrix} 2 & 1 \\ 0 & 0 \end{bmatrix}$$

The system $B\mathbf{x} = \mathbf{0}$ is therefore equivalent to the simpler system

$$\begin{bmatrix} 2 & 1 \\ 0 & 0 \end{bmatrix}\begin{bmatrix} x_1 \\ x_2 \end{bmatrix} = \begin{bmatrix} 0 \\ 0 \end{bmatrix}$$

Since the bottom row of this coefficient matrix contains only zeros, x_2 can be taken as a free variable. The first row then gives

$$2x_1 + x_2 = 0 \quad \Rightarrow \quad x_1 = -\tfrac{1}{2}x_2$$

so any vector of the form

$$\begin{bmatrix} -\tfrac{1}{2}x_2 \\ x_2 \end{bmatrix}$$

satisfies $B\mathbf{x} = \mathbf{0}$. The collection of all such vectors is the nullspace of B, a subspace of $\mathbf{R}^2$:

$$N(B) = \left\{ \begin{bmatrix} -\tfrac{1}{2}x \\ x \end{bmatrix} : x \in \mathbf{R} \right\}$$ ∎

Linear Combinations and the Span of a Collection of Vectors

Let $\mathbf{v}_1, \mathbf{v}_2, \ldots, \mathbf{v}_r$ be vectors in $\mathbf{R}^n$. A **linear combination** of these vectors is any expression of the form

$$k_1\mathbf{v}_1 + k_2\mathbf{v}_2 + \cdots + k_r\mathbf{v}_r$$

where the coefficients $k_1, k_2, \ldots, k_r$ are scalars.

Example 12: The vector $\mathbf{v} = (-7, -6)$ is a linear combination of the vectors $\mathbf{v}_1 = (-2, 3)$ and $\mathbf{v}_2 = (1, 4)$, since $\mathbf{v} = 2\mathbf{v}_1 - 3\mathbf{v}_2$. The zero vector is also a linear combination of $\mathbf{v}_1$ and $\mathbf{v}_2$, since $\mathbf{0} = 0\mathbf{v}_1 + 0\mathbf{v}_2$. In fact, it is easy to see that the zero vector in $\mathbf{R}^n$ is always a linear combination of any collection of vectors $\mathbf{v}_1, \mathbf{v}_2, \ldots, \mathbf{v}_r$ from $\mathbf{R}^n$. ∎

The set of *all* linear combinations of a collection of vectors v_1, v_2, ..., v_r from $\mathbf{R}^n$ is called the **span** of $\{v_1, v_2, ..., v_r\}$. This set, denoted span$\{v_1, v_2, ..., v_r\}$, is always a subspace of $\mathbf{R}^n$, since it is clearly closed under addition and scalar multiplication (because it contains *all* linear combinations of v_1, v_2, ..., v_r). If $V = $ span$\{v_1, v_2, ..., v_r\}$, then V is said to be **spanned** by v_1, v_2, ..., v_r.

Example 13: The span of the set $\{(2, 5, 3), (1, 1, 1)\}$ is the subspace of $\mathbf{R}^3$ consisting of all linear combinations of the vectors $v_1 = (2, 5, 3)$ and $v_2 = (1, 1, 1)$. This defines a plane in $\mathbf{R}^3$. Since a normal vector to this plane is $n = v_1 \times v_2 = (2, 1, -3)$, the equation of this plane has the form $2x + y - 3z = d$ for some constant d. Since the plane must contain the origin—it's a subspace—d must be 0. This is the plane in Example 7. ∎

Example 14: The subspace of $\mathbf{R}^2$ spanned by the vectors $i = (1, 0)$ and $j = (0, 1)$ is all of $\mathbf{R}^2$, because *every* vector in $\mathbf{R}^2$ can be written as a linear combination of i and j:

$$\text{span}\{i, j\} = \mathbf{R}^2 \qquad ∎$$

Let v_1, v_2, ..., v_{r-1}, v_r be vectors in $\mathbf{R}^n$. If v_r is a linear combination of v_1, v_2, ..., v_{r-1}, then

$$\text{span}\{v_1, v_2, ..., v_{r-1}, v_r\} = \text{span}\{v_1, v_2, ..., v_{r-1}\}$$

That is, if any one of the vectors in a given collection is a linear combination of the others, then it can be discarded without affecting the span. Therefore, to arrive at the most "efficient" spanning set, seek out and eliminate any vectors that depend on (that is, can be written as a linear combination of) the others.

Example 15: Let $v_1 = (2, 5, 3)$, $v_2 = (1, 1, 1)$, and $v_3 = (3, 15, 7)$. Since $v_3 = 4v_1 - 5v_2$,

$$\text{span}\{v_1, v_2, v_3\} = \text{span}\{v_1, v_2\}$$

That is, because v_3 is a linear combination of v_1 and v_2, it can be eliminated from the collection without affecting the span. Geometrically, the vector $(3, 15, 7)$ lies in the plane spanned by v_1 and v_2 (see Example 7 above), so adding multiples of v_3 to linear combinations of v_1 and v_2 would yield no vectors off this plane. Note that v_1 is a linear combination of v_2 and v_3 (since $v_1 = 5/4v_2 + 1/4v_3$), and v_2 is a linear combination of v_1 and v_3 (since $v_2 = 4/5v_1 - 1/5v_3$). Therefore, *any one* of these vectors can be discarded without affecting the span:

$$\text{span}\{v_1, v_2, v_3\} = \text{span}\{v_1, v_2\}$$
$$= \text{span}\{v_2, v_3\} = \text{span}\{v_1, v_3\} \quad \blacksquare$$

Example 16: Let $v_1 = (2, 5, 3)$, $v_2 = (1, 1, 1)$, and $v_3 = (4, -2, 0)$. Because there exist no constants k_1 and k_2 such that $v_3 = k_1 v_1 + k_2 v_2$, v_3 is not a linear combination of v_1 and v_2. Therefore, v_3 does not lie in the plane spanned by v_1 and v_2, as shown in Figure 42:

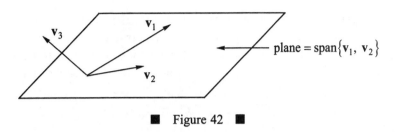

■ Figure 42 ■

Consequently, the span of $\mathbf{v}_1$, $\mathbf{v}_2$, and $\mathbf{v}_3$ contains vectors not in the span of $\mathbf{v}_1$ and $\mathbf{v}_2$ alone. In fact,

$$\text{span}\{\mathbf{v}_1, \ \mathbf{v}_2\} = (\text{the plane } 2x + y - 3z = 0)$$
$$\text{span}\{\mathbf{v}_1, \ \mathbf{v}_2, \ \mathbf{v}_3\} = \text{all of } \mathbf{R}^3 \qquad ■$$

Linear Independence

Let $A = \{\mathbf{v}_1, \mathbf{v}_2, \ldots, \mathbf{v}_r\}$ be a collection of vectors from $\mathbf{R}^n$. If $r \geq 2$ and at least one of the vectors in A can be written as a linear combination of the others, then A is said to be **linearly dependent**. The motivation for this description is simple: At least one of the vectors depends (linearly) on the others. On the other hand, if no vector in A is equal to a linear combination of the others, then A is said to be a **linearly independent** set. It is also quite common to say that "the vectors are linearly dependent (or independent)" rather than "the set containing these vectors is linearly dependent (or independent)."

Example 17: Are the vectors $\mathbf{v}_1 = (2, 5, 3)$, $\mathbf{v}_2 = (1, 1, 1)$, and $\mathbf{v}_3 = (4, -2, 0)$ linearly independent?

If none of these vectors can be expressed as a linear combination of the other two, then the vectors are independent; otherwise, they are dependent. If, for example, v_3 were a linear combination of v_1 and v_2, then there would exist scalars k_1 and k_2 such that $k_1 v_1 + k_2 v_2 = v_3$. This equation reads

$$k_1(2, 5, 3) + k_2(1, 1, 1) = (4, -2, 0)$$

which is equivalent to

$$2k_1 + k_2 = 4$$
$$5k_1 + k_2 = -2$$
$$3k_1 + k_2 = 0$$

However, this is an inconsistent system. For instance, subtracting the first equation from the third yields $k_1 = -4$, and substituting this value into either the first or third equation gives $k_2 = 12$. However, $(k_1, k_2) = (-4, 12)$ does not satisfy the second equation. The conclusion is that v_3 is not a linear combination of v_1 and v_2, as stated in Example 16. A similar argument would show that v_1 is not a linear combination of v_2 and v_3 and that v_2 is not a linear combination of v_1 and v_3. Thus, these three vectors are indeed linearly independent. ■

An alternative—but entirely equivalent and often simpler—definition of linear independence reads as follows. A collection of vectors $v_1, v_2, \ldots, v_r$ from $\mathbf{R}^n$ is linearly independent if the only scalars that satisfy

$$k_1 v_1 + k_2 v_2 + \cdots + k_r v_r = \mathbf{0}$$

are $k_1 = k_2 = \cdots = k_r = 0$. This is called the **trivial** linear combination. If, on the other hand, there exists a *nontrivial* linear combination that gives the zero vector, then the vectors are dependent.

Example 18: Use this second definition to show that the vectors from Example 17—$v_1 = (2, 5, 3)$, $v_2 = (1, 1, 1)$, and $v_3 = (4, -2, 0)$—are linearly independent.

These vectors are linearly independent if the only scalars that satisfy

$$k_1 v_1 + k_2 v_2 + k_3 v_3 = 0 \quad (*)$$

are $k_1 = k_2 = k_3 = 0$. But (*) is equivalent to the homogeneous system

$$\begin{bmatrix} | & | & | \\ v_1 & v_2 & v_3 \\ | & | & | \end{bmatrix} \begin{bmatrix} k_1 \\ k_2 \\ k_3 \end{bmatrix} = 0 \quad (**)$$

Row-reducing the coefficient matrix yields

$$\begin{bmatrix} | & | & | \\ v_1 & v_2 & v_3 \\ | & | & | \end{bmatrix} = \begin{bmatrix} 2 & 1 & 4 \\ 5 & 1 & -2 \\ 3 & 1 & 0 \end{bmatrix} \xrightarrow{-2r_1 \text{ added to } r_2} \begin{bmatrix} 2 & 1 & 4 \\ 1 & -1 & -10 \\ 3 & 1 & 0 \end{bmatrix}$$

$$\xrightarrow{r_1 \leftrightarrow r_2} \begin{bmatrix} 1 & -1 & -10 \\ 2 & 1 & 4 \\ 3 & 1 & 0 \end{bmatrix}$$

$$\xrightarrow[-3r_1 \text{ added to } r_3]{-2r_1 \text{ added to } r_2} \begin{bmatrix} 1 & -1 & -10 \\ 0 & 3 & 24 \\ 0 & 4 & 30 \end{bmatrix}$$

$$\xrightarrow{(-4/3)r_2 \text{ added to } r_3} \begin{bmatrix} 1 & -1 & -10 \\ 0 & 3 & 24 \\ 0 & 0 & -2 \end{bmatrix}$$

This echelon form of the matrix makes it easy to see that $k_3 = 0$, from which follow $k_2 = 0$ and $k_1 = 0$. Thus, equation (**)—and therefore (*)—is satisfied only by $k_1 = k_2 = k_3 = 0$, which proves that the given vectors are linearly independent. ∎

Example 19: Are the vectors $\mathbf{v}_1 = (4, 1, -2)$, $\mathbf{v}_2 = (-3, 0, 1)$, and $\mathbf{v}_3 = (1, -2, 1)$ linearly independent?

The equation $k_1\mathbf{v}_1 + k_2\mathbf{v}_2 + k_3\mathbf{v}_3 = \mathbf{0}$ is equivalent to the homogeneous system

$$\begin{bmatrix} | & | & | \\ \mathbf{v}_1 & \mathbf{v}_2 & \mathbf{v}_3 \\ | & | & | \end{bmatrix} \begin{bmatrix} k_1 \\ k_2 \\ k_3 \end{bmatrix} = \mathbf{0} \quad (*)$$

Row-reduction of the coefficient matrix produces a row of zeros:

$$\begin{bmatrix} | & | & | \\ \mathbf{v}_1 & \mathbf{v}_2 & \mathbf{v}_3 \\ | & | & | \end{bmatrix} = \begin{bmatrix} 4 & -3 & 1 \\ 1 & 0 & -2 \\ -2 & 1 & 1 \end{bmatrix} \xrightarrow{\mathbf{r}_1 \leftrightarrow \mathbf{r}_2} \begin{bmatrix} 1 & 0 & -2 \\ 4 & -3 & 1 \\ -2 & 1 & 1 \end{bmatrix}$$

$$\xrightarrow[\substack{2\mathbf{r}_1 \text{ added to } \mathbf{r}_3}]{-4\mathbf{r}_1 \text{ added to } \mathbf{r}_2} \begin{bmatrix} 1 & 0 & -2 \\ 0 & -3 & 9 \\ 0 & 1 & -3 \end{bmatrix}$$

$$\xrightarrow{(1/3)\mathbf{r}_2 \text{ added to } \mathbf{r}_3} \begin{bmatrix} 1 & 0 & -2 \\ 0 & -3 & 9 \\ 0 & 0 & 0 \end{bmatrix}$$

Since the general solution will contain a free variable, the homogeneous system (*) has nontrivial solutions. This shows that there exists a nontrivial linear combination of the vectors v_1, v_2, and v_3 that gives the zero vector: v_1, v_2, and v_3 are dependent. ■

Example 20: There is exactly one value of c such that the vectors

$$v_1 = (1, 0, 0, 1), \quad v_2 = (0, 1, -1, 0),$$
$$v_3 = (-1, 0, -1, 0), \quad \text{and} \quad v_4 = (1, 1, 1, c)$$

are linearly dependent. Find this value of c and determine a nontrivial linear combination of these vectors that equals the zero vector.

As before, consider the homogeneous system

$$
\begin{bmatrix}
| & | & | & | \\
v_1 & v_2 & v_3 & v_4 \\
| & | & | & |
\end{bmatrix}
\begin{bmatrix}
k_1 \\
k_2 \\
k_3 \\
k_4
\end{bmatrix} = 0
$$

and perform the following elementary row operations on the coefficient matrix:

$$\begin{bmatrix} 1 & 0 & -1 & 1 \\ 0 & 1 & 0 & 1 \\ 0 & -1 & -1 & 1 \\ 1 & 1 & 0 & c \end{bmatrix} \xrightarrow{\ -r_1 \text{ added to } r_4\ } \begin{bmatrix} 1 & 0 & -1 & 1 \\ 0 & 1 & 0 & 1 \\ 0 & -1 & -1 & 1 \\ 0 & 1 & 1 & c-1 \end{bmatrix}$$

$$\xrightarrow[\ -r_2 \text{ added to } r_4\ ]{r_2 \text{ added to } r_3} \begin{bmatrix} 1 & 0 & -1 & 1 \\ 0 & 1 & 0 & 1 \\ 0 & 0 & -1 & 2 \\ 0 & 0 & 1 & c-2 \end{bmatrix}$$

$$\xrightarrow{\ r_3 \text{ added to } r_4\ } \begin{bmatrix} 1 & 0 & -1 & 1 \\ 0 & 1 & 0 & 1 \\ 0 & 0 & -1 & 2 \\ 0 & 0 & 0 & c \end{bmatrix}$$

In order to obtain nontrivial solutions, there must be at least one row of zeros in this echelon form of the matrix. If c is 0, this condition is satisfied. Since $c = 0$, the vector $\mathbf{v}_4$ equals (1, 1, 1, 0). Now, to find a nontrivial linear combination of the vectors $\mathbf{v}_1$, $\mathbf{v}_2$, $\mathbf{v}_3$, and $\mathbf{v}_4$ that gives the zero vector, a particular nontrivial solution to the matrix equation

$$\begin{bmatrix} | & | & | & | \\ \mathbf{v}_1 & \mathbf{v}_2 & \mathbf{v}_3 & \mathbf{v}_4 \\ | & | & | & | \end{bmatrix} \begin{bmatrix} k_1 \\ k_2 \\ k_3 \\ k_4 \end{bmatrix} = \begin{bmatrix} 1 & 0 & -1 & 1 \\ 0 & 1 & 0 & 1 \\ 0 & -1 & -1 & 1 \\ 1 & 1 & 0 & 0 \end{bmatrix} \begin{bmatrix} k_1 \\ k_2 \\ k_3 \\ k_4 \end{bmatrix} = \mathbf{0} \quad (*)$$

is needed. From the row operations performed above, this equation is equivalent to

$$\begin{bmatrix} 1 & 0 & -1 & 1 \\ 0 & 1 & 0 & 1 \\ 0 & 0 & -1 & 2 \\ 0 & 0 & 0 & 0 \end{bmatrix} \begin{bmatrix} k_1 \\ k_2 \\ k_3 \\ k_4 \end{bmatrix} = \mathbf{0} \quad (**)$$

The last row implies that k_4 can be taken as a free variable; let $k_4 = t$. The third row then says

$$-k_3 + 2k_4 = 0 \quad \Rightarrow \quad k_3 = 2k_4 = 2t$$

The second row implies

$$k_2 + k_4 = 0 \quad \Rightarrow \quad k_2 = -k_4 = -t$$

and, finally, the first row gives

$$k_1 - k_3 + k_4 = 0 \quad \Rightarrow \quad k_1 - 2k_4 + k_4 = 0 \quad \Rightarrow \quad k_1 = k_4 = t$$

Thus, the general solution of the homogeneous system $(**)$—and $(*)$—is

$$(k_1, k_2, k_3, k_4)^{\mathrm{T}} = (t, -t, 2t, t)^{\mathrm{T}} \quad (***)$$

for any t in **R**. Choosing $t = 1$, for example, gives $(k_1, k_2, k_3, k_4)^{\mathrm{T}} = (1, -1, 2, 1)^{\mathrm{T}}$, so

$$k_1 \mathbf{v}_1 + k_2 \mathbf{v}_2 + k_3 \mathbf{v}_3 + k_4 \mathbf{v}_4 = \mathbf{v}_1 - \mathbf{v}_2 + 2\mathbf{v}_3 + \mathbf{v}_4$$

is a linear combination of the vectors $\mathbf{v}_1$, $\mathbf{v}_2$, $\mathbf{v}_3$, and $\mathbf{v}_4$ that equals the zero vector. To verify that

$$\mathbf{v}_1 - \mathbf{v}_2 + 2\mathbf{v}_3 + \mathbf{v}_4 = \mathbf{0}$$

simply substitute and simplify:

$$\begin{aligned}
\mathbf{v}_1 - \mathbf{v}_2 + 2\mathbf{v}_3 + \mathbf{v}_4 &= (1,\ 0,\ 0,\ 1) - (0,\ 1,\ -1,\ 1) \\
&\quad + 2(-1,\ 0,\ -1,\ 0) + (1,\ 1,\ 1,\ 0) \\
&= (1-0-2+1,\ 0-1+0+1, \\
&\qquad 0+1-2+1,\ 1-1+0+0) \\
&= (0,\ 0,\ 0,\ 0) \\
&= \mathbf{0} \quad \checkmark
\end{aligned}$$

Infinitely many other nontrivial linear combinations of $\mathbf{v}_1$, $\mathbf{v}_2$, $\mathbf{v}_3$, and $\mathbf{v}_4$ that equal the zero vector can be found by simply choosing any other *nonzero* value of t in (***) and substituting the resulting values of k_1, k_2, k_3, and k_4 in the expression $k_1\mathbf{v}_1 + k_2\mathbf{v}_2 + k_3\mathbf{v}_3 + k_4\mathbf{v}_4$. ∎

If a collection of vectors from $\mathbf{R}^n$ contains more than n vectors, the question of its linear independence is easily answered. If $C = \{\mathbf{v}_1, \mathbf{v}_2, \ldots, \mathbf{v}_m\}$ is a collection of vectors from $\mathbf{R}^n$ and $m > n$, then C must be linearly dependent. To see why this is so, note that the equation

$$k_1\mathbf{v}_1 + k_2\mathbf{v}_2 + \cdots + k_m\mathbf{v}_m = \mathbf{0} \quad (*)$$

is equivalent to the matrix equation

$$\begin{bmatrix} | & | & & | \\ \mathbf{v}_1 & \mathbf{v}_2 & \cdots & \mathbf{v}_m \\ | & | & & | \end{bmatrix} \begin{bmatrix} k_1 \\ k_2 \\ \vdots \\ k_m \end{bmatrix} = \mathbf{0}$$

Since each vector $\mathbf{v}_j$ contains n components, this matrix equation describes a system with m unknowns and n equations. Any homogeneous system with more unknowns than equations has nontrivial solutions (see Theorem B, page 90), a result which applies here since $m > n$. Because equation (*) has nontrivial solutions, the vectors in C cannot be independent.

Example 21: The collection of vectors $\{2\mathbf{i} - \mathbf{j},\ \mathbf{i} + \mathbf{j},\ -\mathbf{i} + 4\mathbf{j}\}$ from $\mathbf{R}^2$ is linearly dependent because *any* collection of 3 (or more) vectors from $\mathbf{R}^2$ must be dependent. Similarly, the collection $\{\mathbf{i} + \mathbf{j} - \mathbf{k}, 2\mathbf{i} - 3\mathbf{j} + \mathbf{k}, \mathbf{i} - 4\mathbf{k}, -2\mathbf{j}, -5\mathbf{i} + \mathbf{j} - 3\mathbf{k}\}$ of vectors from $\mathbf{R}^3$ cannot be independent, because *any* collection of 4 or more vectors from $\mathbf{R}^3$ is dependent. ∎

Example 22: Any collection of vectors from $\mathbf{R}^n$ that contains the zero vector is automatically dependent, for if $\{\mathbf{v}_1, \mathbf{v}_2, \ldots, \mathbf{v}_{r-1}, \mathbf{0}\}$ is such a collection, then for any $k \neq 0$,

$$0\mathbf{v}_1 + 0\mathbf{v}_2 + \cdots + 0\mathbf{v}_{r-1} + k\mathbf{0}$$

is a nontrivial linear combination that gives the zero vector. ∎

The Rank of a Matrix

The maximum number of linearly independent rows in a matrix A is called the **row rank** of A, and the maximum number of linearly independent columns in A is called the **column rank** of A. If A is an m by n matrix, that is, if A has m rows and n columns, then it is obvious that

$$\left. \begin{array}{l} \text{row rank of } A \leq m \\ \text{column rank of } A \leq n \end{array} \right\} \quad (*)$$

What is not so obvious, however, is that for any matrix A,

the row rank of A = the column rank of A

Because of this fact, there is no reason to distinguish between row rank and column rank; the common value is simply called the **rank** of the matrix. Therefore, if A is $m \times n$, it follows from the inequalities in (*) that

$$\text{rank}(A_{m \times n}) \leq \min(m, n) \quad (**)$$

where $\min(m, n)$ denotes the smaller of the two numbers m and n (or their common value if $m = n$). For example, the rank of a 3×5 matrix can be no more than 3, and the rank of a 4×2 matrix can be no more than 2. A 3×5 matrix,

$$\begin{bmatrix} * & * & * & * & * \\ * & * & * & * & * \\ * & * & * & * & * \end{bmatrix}$$

can be thought of as composed of three 5-vectors (the rows) or five 3-vectors (the columns). Although three 5-vectors could be linearly independent, it is not possible to have five 3-vectors that are independent. Any collection of more than three 3-vectors is automatically dependent. Thus, the column rank—and therefore the rank—of such a matrix can be no greater than 3. So, if A is a 3×5 matrix, this argument shows that

$$\text{rank}(A_{3 \times 5}) \leq 3 = \min(3, 5)$$

in accord with (**).

The process by which the rank of a matrix is determined can be illustrated by the following example. Suppose A is the 4×4 matrix

$$\begin{bmatrix} 1 & -2 & 0 & 4 \\ 3 & 1 & 1 & 0 \\ -1 & -5 & -1 & 8 \\ 3 & 8 & 2 & -12 \end{bmatrix}$$

The four row vectors,

$$\mathbf{r_1} = (1, \ -2, \ 0, \ 4)$$
$$\mathbf{r_2} = (3, \ 1, \ 1, \ 0)$$
$$\mathbf{r_3} = (-1, \ -5, \ -1, \ 8)$$
$$\mathbf{r_4} = (3, \ 8, \ 2, \ -12)$$

are not independent, since, for example,

$$\mathbf{r_3} = 2\mathbf{r_1} - \mathbf{r_2} \quad \text{and} \quad \mathbf{r_4} = -3\mathbf{r_1} + 2\mathbf{r_2} \quad (***)$$

The fact that the vectors $\mathbf{r_3}$ and $\mathbf{r_4}$ can be written as linear combinations of the other two ($\mathbf{r_1}$ and $\mathbf{r_2}$, which are independent) means that the maximum number of independent rows is 2. Thus, the row rank—and therefore the rank—of this matrix is 2.

The equations in (***) can be rewritten as follows:

$$-2\mathbf{r_1} + \mathbf{r_2} + \mathbf{r_3} = \mathbf{0} \quad \text{and} \quad 3\mathbf{r_1} - 2\mathbf{r_2} + \mathbf{r_4} = \mathbf{0}$$

The first equation here implies that if -2 times the first row is added to the third and then the second row is added to the (new) third row, the third row will be become $\mathbf{0}$, a row of zeros. The second equation above says that similar operations performed on the fourth row can produce a row of zeros there also. If after these operations are completed, -3 times the first row is then added to the second row (to clear out all entries below the entry $a_{11} = 1$ in the first column), these elementary row operations reduce the original matrix A to the echelon form

$$\begin{bmatrix} 1 & -2 & 0 & 4 \\ 0 & 7 & 1 & -12 \\ 0 & 0 & 0 & 0 \\ 0 & 0 & 0 & 0 \end{bmatrix}$$

The fact that there are exactly 2 nonzero rows in the reduced form of the matrix indicates that the maximum number of linearly independent rows is 2; hence, rank $A = 2$, in agreement with the conclusion above. In general, then, *to compute the rank of a matrix, perform elementary row operations until the matrix is left in echelon form; the number of nonzero rows remaining in the reduced matrix is the rank.* [Note: Since column rank = row rank, only two of the four *columns* in A—c_1, c_2, c_3, and c_4—are linearly independent. Show that this is indeed the case by verifying the relations

$$c_2 = -2c_1 + 7c_3 \quad \text{and} \quad c_4 = 4c_1 - 12c_3$$

(and checking that c_1 and c_3 are independent). The reduced form of A makes these relations especially easy to see.]

Example 23: Find the rank of the matrix

$$B = \begin{bmatrix} 2 & -1 & 3 \\ 1 & 0 & 1 \\ 0 & 2 & -1 \\ 1 & 1 & 4 \end{bmatrix}$$

First, because the matrix is 4×3, its rank can be no greater than 3. Therefore, at least one of the four rows will become a row of zeros. Perform the following row operations:

$$
\begin{bmatrix}
2 & -1 & 3 \\
1 & 0 & 1 \\
0 & 2 & -1 \\
1 & 1 & 4
\end{bmatrix}
\xrightarrow{\ r_1 \leftrightarrow r_2\ }
\begin{bmatrix}
1 & 0 & 1 \\
2 & -1 & 3 \\
0 & 2 & -1 \\
1 & 1 & 4
\end{bmatrix}
$$

$$
\xrightarrow[\substack{-r_1 \text{ added to } r_4}]{-2r_1 \text{ added to } r_2}
\begin{bmatrix}
1 & 0 & 1 \\
0 & -1 & 1 \\
0 & 2 & -1 \\
0 & 1 & 3
\end{bmatrix}
$$

$$
\xrightarrow[\substack{r_2 \text{ added to } r_4}]{2r_2 \text{ added to } r_3}
\begin{bmatrix}
1 & 0 & 1 \\
0 & -1 & 1 \\
0 & 0 & 1 \\
0 & 0 & 4
\end{bmatrix}
$$

$$
\xrightarrow[\substack{-4r_3 \text{ added to } r_4}]{(-1)r_2}
\begin{bmatrix}
1 & 0 & 1 \\
0 & 1 & -1 \\
0 & 0 & 1 \\
0 & 0 & 0
\end{bmatrix}
$$

Since there are 3 nonzero rows remaining in this echelon form of B,

$$\text{rank } B = 3 \qquad \blacksquare$$

Example 24: Determine the rank of the 4 by 4 checkerboard matrix

$$C = \begin{bmatrix} 1 & -1 & 1 & -1 \\ -1 & 1 & -1 & 1 \\ 1 & -1 & 1 & -1 \\ -1 & 1 & -1 & 1 \end{bmatrix}$$

Since $r_2 = r_4 = -r_1$ and $r_3 = r_1$, all rows but the first vanish upon row-reduction:

$$\begin{bmatrix} 1 & -1 & 1 & -1 \\ -1 & 1 & -1 & 1 \\ 1 & -1 & 1 & -1 \\ -1 & 1 & -1 & 1 \end{bmatrix} \xrightarrow[\substack{-r_1 \text{ added to } r_3 \\ r_1 \text{ added to } r_4}]{r_1 \text{ added to } r_2} \begin{bmatrix} 1 & -1 & 1 & -1 \\ 0 & 0 & 0 & 0 \\ 0 & 0 & 0 & 0 \\ 0 & 0 & 0 & 0 \end{bmatrix}$$

Since only 1 nonzero row remains, rank $C = 1$. ∎

A Basis for a Vector Space

Let V be a subspace of $\mathbf{R}^n$ for some n. A collection $B = \{v_1, v_2, \ldots, v_r\}$ of vectors from V is said to be a **basis** for V if B is linearly independent and spans V. If either one of these criteria is not satisfied, then the collection is not a basis for V. If a collection of vectors spans V, then it contains enough vectors so that every vector in V can be written as a linear combination of those in the collection. If the collection is linearly independent, then it doesn't contain so many vectors that some become dependent on the others. Intuitively, then, a basis has just the right size: It's big enough to span the space but not so big as to be dependent.

Example 25: The collection $\{\mathbf{i}, \mathbf{j}\}$ is a basis for $\mathbf{R}^2$, since it spans $\mathbf{R}^2$ (Example 14) and the vectors $\mathbf{i}$ and $\mathbf{j}$ are linearly independent (because neither is a multiple of the other). This is called the **standard basis** for $\mathbf{R}^2$. Similarly, the set $\{\mathbf{i}, \mathbf{j}, \mathbf{k}\}$ is called the standard basis for $\mathbf{R}^3$, and, in general,

$$\{\mathbf{e}_1 = (1, 0, 0, \ldots, 0), \mathbf{e}_2 = (0, 1, 0, \ldots, 0), \ldots,$$
$$\mathbf{e}_n = (0, 0, \ldots, 0, 1)\}$$

is the standard basis for $\mathbf{R}^n$. ■

Example 26: The collection $\{\mathbf{i}, \mathbf{i}+\mathbf{j}, 2\mathbf{j}\}$ is not a basis for $\mathbf{R}^2$. Although it spans $\mathbf{R}^2$, it is not linearly independent. No collection of 3 or more vectors from $\mathbf{R}^2$ can be independent. ■

Example 27: The collection $\{\mathbf{i}+\mathbf{j}, \mathbf{j}+\mathbf{k}\}$ is not a basis for $\mathbf{R}^3$. Although it is linearly independent, it does not span all of $\mathbf{R}^3$. For example, there exists no linear combination of $\mathbf{i}+\mathbf{j}$ and $\mathbf{j}+\mathbf{k}$ that equals $\mathbf{i}+\mathbf{j}+\mathbf{k}$. ■

Example 28: The collection $\{\mathbf{i}+\mathbf{j}, \mathbf{i}-\mathbf{j}\}$ is a basis for $\mathbf{R}^2$. First, it is linearly independent, since neither $\mathbf{i}+\mathbf{j}$ nor $\mathbf{i}-\mathbf{j}$ is a multiple of the other. Second, it spans all of $\mathbf{R}^2$ because every vector in $\mathbf{R}^2$ can be expressed as a linear combination of $\mathbf{i}+\mathbf{j}$ and $\mathbf{i}-\mathbf{j}$. Specifically, if $a\mathbf{i}+b\mathbf{j}$ is any vector in $\mathbf{R}^2$, then

$$k_1(\mathbf{i}+\mathbf{j})+k_2(\mathbf{i}-\mathbf{j}) = a\mathbf{i}+b\mathbf{j}$$

if $k_1 = \frac{1}{2}(a+b)$ and $k_2 = \frac{1}{2}(a-b)$. ■

Examples 25 and 28 showed that a space may have many different bases. For example, both $\{\mathbf{i}, \mathbf{j}\}$ and $\{\mathbf{i}+\mathbf{j}, \mathbf{i}-\mathbf{j}\}$ are bases for $\mathbf{R}^2$. In fact, *any* collection containing exactly two linearly independent vectors from $\mathbf{R}^2$ is a basis for $\mathbf{R}^2$. Similarly, any collection containing exactly three linearly independent vectors from $\mathbf{R}^3$ is a basis for $\mathbf{R}^3$, and so on. Although no nontrivial subspace of $\mathbf{R}^n$ has a unique basis, there *is* something that all bases for a given space must have in common.

Let V be a subspace of $\mathbf{R}^n$ for some n. If V has a basis containing exactly r vectors, then *every* basis for V contains exactly r vectors. That is, the choice of basis vectors for a given space is not unique, but the *number* of basis vectors *is* unique. This fact permits the following notion to be well defined: The number of vectors in a basis for a vector space $V \subseteq \mathbf{R}^n$ is called the **dimension** of V, denoted dim V.

Example 29: Since the standard basis for $\mathbf{R}^2$, $\{\mathbf{i}, \mathbf{j}\}$, contains exactly 2 vectors, *every* basis for $\mathbf{R}^2$ contains exactly 2 vectors, so dim $\mathbf{R}^2 = 2$. Similarly, since $\{\mathbf{i}, \mathbf{j}, \mathbf{k}\}$ is a basis for $\mathbf{R}^3$ that contains exactly 3 vectors, every basis for $\mathbf{R}^3$ contains exactly 3 vectors, so dim $\mathbf{R}^3 = 3$. In general, dim $\mathbf{R}^n = n$ for every natural number n. ■

Example 30: In $\mathbf{R}^3$, the vectors $\mathbf{i}$ and $\mathbf{k}$ span a subspace of dimension 2. It is the x-z plane, as shown in Figure 43.

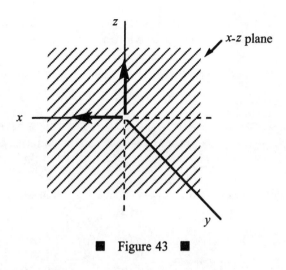

■ Figure 43 ■

Example 31: The one-element collection $\{\mathbf{i} + \mathbf{j} = (1,\ 1)\}$ is a basis for the 1-dimensional subspace V of $\mathbf{R}^2$ consisting of the line $y = x$. See Figure 44.

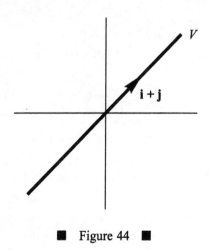

■ Figure 44 ■

Example 32: The trivial subspace, $\{\mathbf{0}\}$, of $\mathbf{R}^n$ is said to have dimension 0. To be consistent with the definition of dimension, then, a basis for $\{\mathbf{0}\}$ must be a collection containing zero elements; this is the empty set, $\varnothing$. ∎

The subspaces of $\mathbf{R}^1$, $\mathbf{R}^2$, and $\mathbf{R}^3$, some of which have been illustrated in the preceding examples, can be summarized as follows:

	subspaces of $\mathbf{R}^1$	subspaces of $\mathbf{R}^2$	subspaces of $\mathbf{R}^3$
dim = 0:	$\{\mathbf{0}\}$	$\{\mathbf{0}\}$	$\{\mathbf{0}\}$
dim = 1:	$\mathbf{R}^1$	lines through the origin	lines through the origin
dim = 2:		$\mathbf{R}^2$	planes through the origin
dim = 3:			$\mathbf{R}^3$

Example 33: Find the dimension of the subspace V of $\mathbf{R}^4$ spanned by the vectors

$$\mathbf{v}_1 = (1, \ -2, \ 0, \ 4)$$
$$\mathbf{v}_2 = (3, \ 1, \ 1, \ 0)$$
$$\mathbf{v}_3 = (-1, \ -5, \ -1, \ 8)$$
$$\mathbf{v}_4 = (3, \ 8, \ 2, \ -12)$$

The collection $\{\mathbf{v}_1, \mathbf{v}_2, \mathbf{v}_3, \mathbf{v}_4\}$ is not a basis for V—and dim V is not 4—because $\{\mathbf{v}_1, \mathbf{v}_2, \mathbf{v}_3, \mathbf{v}_4\}$ is not linearly independent; see the calculation preceding Example 23 above. Discarding $\mathbf{v}_3$ and $\mathbf{v}_4$ from this collection does not diminish the span of

$\{\mathbf{v}_1, \mathbf{v}_2, \mathbf{v}_3, \mathbf{v}_4\}$, but the resulting collection, $\{\mathbf{v}_1, \mathbf{v}_2\}$, is linearly independent. Thus, $\{\mathbf{v}_1, \mathbf{v}_2\}$ is a basis for V, so dim $V = 2$. ■

Example 34: Find the dimension of the span of the vectors

$$\mathbf{w}_1 = (1, 2, 3, 4, 5), \quad \mathbf{w}_2 = (-2, 1, -3, -5, -4),$$

$$\text{and} \quad \mathbf{w}_3 = (-1, 8, 3, 2, 7)$$

Since these vectors are in $\mathbf{R}^5$, their span, S, is a subspace of $\mathbf{R}^5$. It is not, however, a 3-dimensional subspace of $\mathbf{R}^5$, since the three vectors $\mathbf{w}_1$, $\mathbf{w}_2$, and $\mathbf{w}_3$ are not linearly independent. In fact, since $\mathbf{w}_3 = 3\mathbf{w}_1 + 2\mathbf{w}_2$, the vector $\mathbf{w}_3$ can be discarded from the collection without diminishing the span. Since the vectors $\mathbf{w}_1$ and $\mathbf{w}_2$ are independent—neither is a scalar multiple of the other—the collection $\{\mathbf{w}_1, \mathbf{w}_2\}$ serves as a basis for S, so its dimension is 2. ■

The most important attribute of a basis is the ability to write every vector in the space in a *unique* way in terms of the basis vectors. To see why this is so, let $B = \{\mathbf{v}_1, \mathbf{v}_2, \ldots, \mathbf{v}_r\}$ be a basis for a vector space V. Since a basis must span V, every vector $\mathbf{v}$ in V can be written in at least one way as a linear combination of the vectors in B. That is, there exist scalars k_1, $k_2, \ldots, k_r$ such that

$$k_1\mathbf{v}_1 + k_2\mathbf{v}_2 + \cdots + k_r\mathbf{v}_r = \mathbf{v} \quad (*)$$

To show that no other choice of scalar multiples could give $\mathbf{v}$, assume that

$$k_1'\mathbf{v}_1 + k_2'\mathbf{v}_2 + \cdots + k_r'\mathbf{v}_r = \mathbf{v} \quad (**)$$

is also a linear combination of the basis vectors that equals $\mathbf{v}$.

Subtracting (*) from (**) yields

$$(k_1' - k_1)\mathbf{v}_1 + (k_2' - k_2)\mathbf{v}_2 + \cdots + (k_r' - k_r)\mathbf{v}_r = \mathbf{0} \quad (***)$$

This expression is a linear combination of the basis vectors that gives the zero vector. Since the basis vectors must be linearly independent, each of the scalars in (***) must be zero:

$$k_1' - k_1 = 0, \quad k_2' - k_2 = 0, \quad \ldots, \quad k_r' - k_r = 0$$

Therefore, $k_1' = k_1$, $k_2' = k_2$, ..., and $k_r' = k_r$, so the representation in (*) is indeed unique. When $\mathbf{v}$ is written as the linear combination (*) of the basis vectors $\mathbf{v}_1$, $\mathbf{v}_2$, . . ., $\mathbf{v}_r$, the uniquely determined scalar coefficients $k_1, k_2, \ldots, k_r$ are called the **components** of $\mathbf{v}$ relative to the basis B. The row vector $(k_1, k_2, \ldots, k_r)$ is called the **component vector** of $\mathbf{v}$ relative to B and is denoted $(\mathbf{v})_B$. Sometimes, it is convenient to write the component vector as a *column* vector; in this case, the component vector $(k_1, k_2, \ldots, k_r)^{\mathrm{T}}$ is denoted $[\mathbf{v}]_B$.

Example 35: Consider the collection $C = \{\mathbf{i}, \mathbf{i} + \mathbf{j}, 2\mathbf{j}\}$ of vectors in $\mathbf{R}^2$. Note that the vector $\mathbf{v} = 3\mathbf{i} + 4\mathbf{j}$ can be written as a linear combination of the vectors in C as follows:

$$3\mathbf{i} + 4\mathbf{j} = 1(\mathbf{i}) + 2(\mathbf{i} + \mathbf{j}) + 1(2\mathbf{j})$$

and

$$3\mathbf{i} + 4\mathbf{j} = 3(\mathbf{i}) + 0(\mathbf{i} + \mathbf{j}) + 2(2\mathbf{j})$$

The fact that there is more than one way to express the vector $\mathbf{v}$ in $\mathbf{R}^2$ as a linear combination of the vectors in C provides another indication (besides the simpler one stated in Example 26) that C cannot be a basis for $\mathbf{R}^2$. If C were a basis, the vector $\mathbf{v}$ could be written as a linear combination of the vectors in C in one *and only one* way. ∎

Example 36: Consider the basis $B = \{i + j, \ 2i - j\}$ of $\mathbf{R}^2$. Determine the components of the vector $v = 2i - 7j$ relative to B.

The components of v relative to B are the scalar coefficients k_1 and k_2 which satisfy the equation

$$k_1(i + j) + k_2(2i - j) = 2i - 7j$$

This equation is equivalent to the system

$$k_1 + 2k_2 = 2$$
$$k_1 - k_2 = -7$$

The solution to this system is $k_1 = -4$ and $k_2 = 3$, so

$$(v)_B = (-4, 3) \quad \blacksquare$$

Example 37: Relative to the standard basis $\{i, \ j, \ k\} = \{\hat{e}_1, \ \hat{e}_2, \ \hat{e}_3\}$ for $\mathbf{R}^3$, the component vector of any vector v in $\mathbf{R}^3$ is equal to v itself: $(v)_B = v$. This same result holds for the standard basis $\{\hat{e}_1, \ \hat{e}_2, \ \ldots, \ \hat{e}_n\}$ for every $\mathbf{R}^n$. $\quad \blacksquare$

Orthonormal bases. If $B = \{v_1, \ v_2, \ \ldots, \ v_n\}$ is a basis for a vector space V, then every vector v in V can be written as a linear combination of the basis vectors in one and only one way:

$$k_1 v_1 + k_2 v_2 + \cdots + k_n v_n = v$$

Finding the components of v relative to the basis B—the scalar coefficients $k_1, \ k_2, \ \ldots, \ k_n$ in the representation above—generally involves solving a system of equations, as in Example 36 above. However, if the basis vectors are **orthonormal**, that is, mutually orthogonal unit vectors, then the calculation of the components is especially easy. Here's why. Assume that $B = \{\hat{\mathbf{v}}_1, \hat{\mathbf{v}}_2, \ldots, \hat{\mathbf{v}}_n\}$ is an orthonormal basis. Starting with the equation above—with $\hat{\mathbf{v}}_1, \hat{\mathbf{v}}_2, \ldots, \hat{\mathbf{v}}_n$ replacing $\mathbf{v}_1, \mathbf{v}_2, \ldots, \mathbf{v}_n$ to emphasize that the basis vectors are now assumed to be unit vectors—take the dot product of both sides with $\hat{\mathbf{v}}_1$:

$$(k_1\hat{\mathbf{v}}_1 + k_2\hat{\mathbf{v}}_2 + \cdots + k_n\hat{\mathbf{v}}_n) \cdot \hat{\mathbf{v}}_1 = \mathbf{v} \cdot \hat{\mathbf{v}}_1$$

By the linearity of the dot product, the left-hand side becomes

$$k_1(\hat{\mathbf{v}}_1 \cdot \hat{\mathbf{v}}_1) + k_2(\hat{\mathbf{v}}_2 \cdot \hat{\mathbf{v}}_1) + \cdots + k_n(\hat{\mathbf{v}}_n \cdot \hat{\mathbf{v}}_1) = \mathbf{v} \cdot \hat{\mathbf{v}}_1$$

Now, by the orthogonality of the basis vectors, $\hat{\mathbf{v}}_i \cdot \hat{\mathbf{v}}_1 = 0$ for $i = 2$ through n. Furthermore, because $\hat{\mathbf{v}}_1$ is a unit vector, $\hat{\mathbf{v}}_1 \cdot \hat{\mathbf{v}}_1 = \|\hat{\mathbf{v}}_1\|^2 = 1^2 = 1$. Therefore, the equation above simplifies to the statement

$$k_1 = \mathbf{v} \cdot \hat{\mathbf{v}}_1$$

In general, if $B = \{\hat{\mathbf{v}}_1, \hat{\mathbf{v}}_2, \ldots, \hat{\mathbf{v}}_n\}$ is an orthonormal basis for a vector space V, then the components, k_i, of any vector $\mathbf{v}$ relative to B are found from the simple formula

$$k_i = \text{component } i \text{ of } (\mathbf{v})_B = \mathbf{v} \cdot \hat{\mathbf{v}}_i$$

Example 38: Consider the vectors

$$\mathbf{v}_1 = (-2, 2, 1), \quad \mathbf{v}_2 = (1, -1, 4), \quad \text{and} \quad \mathbf{v}_3 = (1, 1, 0)$$

from $\mathbf{R}^3$. These vectors are mutually orthogonal, as you may easily verify by checking that $\mathbf{v}_1 \cdot \mathbf{v}_2 = \mathbf{v}_1 \cdot \mathbf{v}_3 = \mathbf{v}_2 \cdot \mathbf{v}_3 = 0$. Normalize these vectors, thereby obtaining an orthonormal basis for $\mathbf{R}^3$ and then find the components of the vector $\mathbf{v} = (1, 2, 3)$ relative to this basis.

A nonzero vector is *normalized*—made into a unit vector—by dividing it by its length. Therefore,

$$\hat{\mathbf{v}}_1 = \frac{\mathbf{v}_1}{\|\mathbf{v}_1\|} = \frac{(-2,\ 2,\ 1)}{3} = \left(-\tfrac{2}{3},\ \tfrac{2}{3},\ \tfrac{1}{3}\right)$$

$$\hat{\mathbf{v}}_2 = \frac{\mathbf{v}_2}{\|\mathbf{v}_2\|} = \frac{(1,\ -1,\ 4)}{\sqrt{18}} = \left(\tfrac{1}{3\sqrt{2}},\ -\tfrac{1}{3\sqrt{2}},\ \tfrac{4}{3\sqrt{2}}\right)$$

$$\hat{\mathbf{v}}_3 = \frac{\mathbf{v}_3}{\|\mathbf{v}_3\|} = \frac{(1,\ 1,\ 0)}{\sqrt{2}} = \left(\tfrac{1}{\sqrt{2}},\ \tfrac{1}{\sqrt{2}},\ 0\right)$$

Since $B = \{\hat{\mathbf{v}}_1,\ \hat{\mathbf{v}}_2,\ \hat{\mathbf{v}}_3\}$ is an orthonormal basis for $\mathbf{R}^3$, the result stated above guarantees that the components of $\mathbf{v}$ relative to B are found by simply taking the following dot products:

$$k_1 = \mathbf{v} \cdot \hat{\mathbf{v}}_1 = (1,\ 2,\ 3) \cdot \left(-\tfrac{2}{3},\ \tfrac{2}{3},\ \tfrac{1}{3}\right) = \tfrac{5}{3}$$

$$k_2 = \mathbf{v} \cdot \hat{\mathbf{v}}_2 = (1,\ 2,\ 3) \cdot \left(\tfrac{1}{3\sqrt{2}},\ -\tfrac{1}{3\sqrt{2}},\ \tfrac{4}{3\sqrt{2}}\right) = \tfrac{11}{3\sqrt{2}}$$

$$k_3 = \mathbf{v} \cdot \hat{\mathbf{v}}_3 = (1,\ 2,\ 3) \cdot \left(\tfrac{1}{\sqrt{2}},\ \tfrac{1}{\sqrt{2}},\ 0\right) = \tfrac{3}{\sqrt{2}}$$

Therefore, $(\mathbf{v})_B = (5/3,\ 11/(3\sqrt{2}),\ 3/\sqrt{2})$, which means that the unique representation of $\mathbf{v}$ as a linear combination of the basis vectors reads $\mathbf{v} = 5/3\,\hat{\mathbf{v}}_1 + 11/(3\sqrt{2})\,\hat{\mathbf{v}}_2 + 3/\sqrt{2}\,\hat{\mathbf{v}}_3$, as you may verify. ∎

Example 39: Prove that a set of mutually orthogonal, nonzero vectors is linearly independent.

Proof. Let $\{\mathbf{v}_1,\ \mathbf{v}_2,\ \ldots,\ \mathbf{v}_r\}$ be a set of nonzero vectors from some $\mathbf{R}^n$ which are mutually orthogonal, which means that no $\mathbf{v}_i = \mathbf{0}$ and $\mathbf{v}_i \cdot \mathbf{v}_j = 0$ for $i \neq j$. Let

$$k_1\mathbf{v}_1 + k_2\mathbf{v}_2 + \cdots + k_r\mathbf{v}_r = \mathbf{0} \qquad (*)$$

be a linear combination of the vectors in this set that gives the zero vector. The goal is to show that $k_1 = k_2 = \cdots = k_r = 0$. To this end, take the dot product of both sides of the equation with $\mathbf{v}_1$:

$$(k_1\mathbf{v}_1 + k_2\mathbf{v}_2 + \cdots + k_r\mathbf{v}_r) \cdot \mathbf{v}_1 = \mathbf{0} \cdot \mathbf{v}_1$$
$$k_1(\mathbf{v}_1 \cdot \mathbf{v}_1) + k_2(\mathbf{v}_2 \cdot \mathbf{v}_1) + \cdots + k_n(\mathbf{v}_r \cdot \mathbf{v}_1) = 0$$
$$k_1\|\mathbf{v}_1\|^2 + k_2(0) + \cdots + k_r(0) = 0$$
$$k_1 = 0$$

The second equation follows from the first by the linearity of the dot product, the third equation follows from the second by the orthogonality of the vectors, and the final equation is a consequence of the fact that $\|\mathbf{v}_1\|^2 \neq 0$ (since $\mathbf{v}_1 \neq \mathbf{0}$). It is now easy to see that taking the dot product of both sides of (*) with $\mathbf{v}_i$ yields $k_i = 0$, establishing that *every* scalar coefficient in (*) must be zero, thus confirming that the vectors $\mathbf{v}_1, \mathbf{v}_2, \ldots, \mathbf{v}_r$ are indeed independent. ∎

Projection onto a Subspace

Let S be a nontrivial subspace of a vector space V and assume that $\mathbf{v}$ is a vector in V that does not lie in S. Then the vector $\mathbf{v}$ can be uniquely written as a sum, $\mathbf{v}_{\|S} + \mathbf{v}_{\perp S}$, where $\mathbf{v}_{\|S}$ is parallel to S and $\mathbf{v}_{\perp S}$ is orthogonal to S; see Figure 45.

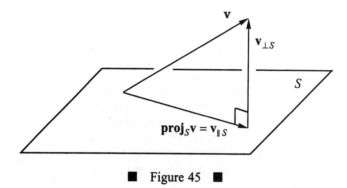

■ Figure 45 ■

The vector $\mathbf{v}_{\|S}$, which actually lies *in* S, is called the **projection** of $\mathbf{v}$ onto S, also denoted $\mathbf{proj}_S\mathbf{v}$. If $\mathbf{v}_1, \mathbf{v}_2, \ldots, \mathbf{v}_r$ form an *orthogonal* basis for S, then the projection of $\mathbf{v}$ onto S is the sum of the projections of $\mathbf{v}$ onto the individual basis vectors, a fact that depends critically on the basis vectors being orthogonal:

$$\mathbf{proj}_S\mathbf{v} = \mathbf{proj}_{\mathbf{v}_1}\mathbf{v} + \mathbf{proj}_{\mathbf{v}_2}\mathbf{v} + \cdots + \mathbf{proj}_{\mathbf{v}_r}\mathbf{v}$$

$$= \frac{\mathbf{v}\cdot\mathbf{v}_1}{\mathbf{v}_1\cdot\mathbf{v}_1}\mathbf{v}_1 + \frac{\mathbf{v}\cdot\mathbf{v}_2}{\mathbf{v}_2\cdot\mathbf{v}_2}\mathbf{v}_2 + \cdots + \frac{\mathbf{v}\cdot\mathbf{v}_r}{\mathbf{v}_r\cdot\mathbf{v}_r}\mathbf{v}_r \quad (*)$$

Figure 46 shows geometrically why this formula is true in the case of a 2-dimensional subspace S in $\mathbf{R}^3$.

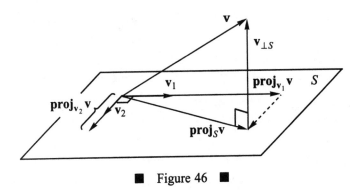

■ Figure 46 ■

Example 40: Let S be the 2-dimensional subspace of $\mathbf{R}^3$ spanned by the orthogonal vectors $\mathbf{v}_1 = (1, 2, 1)$ and $\mathbf{v}_2 = (1, -1, 1)$. Write the vector $\mathbf{v} = (-2, 2, 2)$ as the sum of a vector in S and a vector orthogonal to S.

From (*), the projection of $\mathbf{v}$ onto S is the vector

$$\mathbf{proj}_S\mathbf{v} = \mathbf{proj}_{\mathbf{v}_1}\mathbf{v} + \mathbf{proj}_{\mathbf{v}_2}\mathbf{v}$$

$$= \frac{\mathbf{v}\cdot\mathbf{v}_1}{\mathbf{v}_1\cdot\mathbf{v}_1}\mathbf{v}_1 + \frac{\mathbf{v}\cdot\mathbf{v}_2}{\mathbf{v}_2\cdot\mathbf{v}_2}\mathbf{v}_2$$

$$= \frac{(-2)(1)+(2)(2)+(2)(1)}{(1)(1)+(2)(2)+(1)(1)}(1,\ 2,\ 1)$$

$$\quad + \frac{(-2)(1)+(2)(-1)+(2)(1)}{(1)(1)+(-1)(-1)+(1)(1)}(1,\ -1,\ 1)$$

$$= \tfrac{4}{6}(1,\ 2,\ 1) - \tfrac{2}{3}(1,\ -1,\ 1)$$

$$= (0,\ 2,\ 0)$$

Therefore, $\mathbf{v} = \mathbf{v}_{\|S} + \mathbf{v}_{\perp S}$, where $\mathbf{v}_{\|S} = \mathbf{proj}_S\mathbf{v} = (0,\ 2,\ 0)$ and

$$\mathbf{v}_{\perp S} = \mathbf{v} - \mathbf{v}_{\parallel S}$$
$$= (-2,\ 2,\ 2) - (0,\ 2,\ 0)$$
$$= (-2,\ 0,\ 2)$$

That $\mathbf{v}_{\perp S} = (-2,\ 0,\ 2)$ truly is orthogonal to S is proved by noting that it is orthogonal to both $\mathbf{v}_1$ and $\mathbf{v}_2$:

$$\mathbf{v}_{\perp S} \cdot \mathbf{v}_1 = (-2,\ 0,\ 2) \cdot (1,\ 2,\ 1) = 0$$
$$\mathbf{v}_{\perp S} \cdot \mathbf{v}_2 = (-2,\ 0,\ 2) \cdot (1,\ -1,\ 1) = 0$$

In summary, then, the unique representation of the vector $\mathbf{v}$ as the sum of a vector in S and a vector orthogonal to S reads as follows:

$$\underbrace{(-2,\ 2,\ 2)}_{\mathbf{v}} = \underbrace{(0,\ 2,\ 0)}_{\mathbf{v}_{\parallel S}} + \underbrace{(-2,\ 0,\ 2)}_{\mathbf{v}_{\perp S}}$$

See Figure 47.

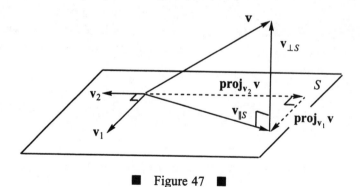

■ Figure 47 ■

Example 41: Let S be a subspace of a Euclidean vector space V. The collection of all vectors in V that are orthogonal to every vector in S is called the **orthogonal complement** of S:

$$S^\perp = \left\{ \mathbf{v} \in V : \mathbf{v} \perp \mathbf{s} \text{ for every } \mathbf{s} \in S \right\}$$
$$= \left\{ \mathbf{v} \in V : \mathbf{v} \cdot \mathbf{s} = 0 \text{ for every } \mathbf{s} \in S \right\}$$

($S^\perp$ is read "S perp.") Show that $S^\perp$ is also a subspace of V.

Proof. First, note that $S^\perp$ is nonempty, since $\mathbf{0} \in S^\perp$. In order to prove that $S^\perp$ is a subspace, closure under vector addition and scalar multiplication must be established. Let $\mathbf{v}_1$ and $\mathbf{v}_2$ be vectors in $S^\perp$; since $\mathbf{v}_1 \cdot \mathbf{s} = \mathbf{v}_2 \cdot \mathbf{s} = 0$ for every vector $\mathbf{s}$ in S,

$$(\mathbf{v}_1 + \mathbf{v}_2) \cdot \mathbf{s} = \mathbf{v}_1 \cdot \mathbf{s} + \mathbf{v}_2 \cdot \mathbf{s}$$
$$= 0 + 0$$
$$= 0 \quad \text{for every } \mathbf{s} \in S$$

proving that $\mathbf{v}_1 + \mathbf{v}_2 \in S^\perp$. Therefore, $S^\perp$ is closed under vector addition. Finally, if k is a scalar, then for any $\mathbf{v}$ in $S^\perp$, $(k\mathbf{v}) \cdot \mathbf{s} = k(\mathbf{v} \cdot \mathbf{s}) = k(0) = 0$ for every vector $\mathbf{s}$ in S, which shows that $S^\perp$ is also closed under scalar multiplication. This completes the proof. ∎

Example 42: Find the orthogonal complement of the x-y plane in $\mathbf{R}^3$.

At first glance, it might seem that the x-z plane is the orthogonal complement of the x-y plane, just as a wall is perpendicular to the floor. However, not every vector in the x-z plane is orthogonal to every vector in the x-y plane: for example, the vector $\mathbf{v} = (1, 0, 1)$ in the x-z plane is not orthogonal to the vector $\mathbf{w} = (1, 1, 0)$ in the x-y plane, since $\mathbf{v} \cdot \mathbf{w} = 1 \neq 0$. See Figure 48. The vectors that are orthogonal to every vector in the x-y plane are only those along the z axis; *this* is the orthogonal complement in $\mathbf{R}^3$ of the x-y plane. In fact, it can be shown that if S is a k-dimensional subspace of $\mathbf{R}^n$, then dim $S^\perp = n - k$; thus, dim S + dim $S^\perp = n$, the dimension of the entire space. Since the x-y plane is a 2-dimensional subspace of $\mathbf{R}^3$, its orthogonal complement in $\mathbf{R}^3$ must have dimension $3 - 2 = 1$. This result would remove the x-z plane, which is 2-dimensional, from consideration as the orthogonal complement of the x-y plane.

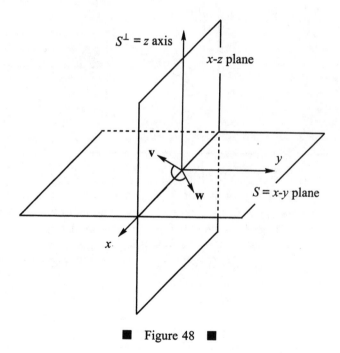

■ Figure 48 ■

Example 43: Let P be the subspace of $\mathbf{R}^3$ specified by the equation $2x + y - 2z = 0$. Find the distance between P and the point $\mathbf{q} = (3, 2, 1)$.

The subspace P is clearly a plane in $\mathbf{R}^3$, and $\mathbf{q}$ is a point that does not lie in P. From Figure 49, it is clear that the distance from $\mathbf{q}$ to P is the length of the component of $\mathbf{q}$ orthogonal to P.

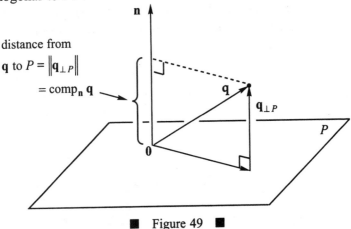

distance from
$\mathbf{q}$ to $P = \|\mathbf{q}_{\perp P}\|$
$= \text{comp}_{\mathbf{n}}\, \mathbf{q}$

■ Figure 49 ■

One way to find the orthogonal component $\mathbf{q}_{\perp P}$ is to find an orthogonal basis for P, use these vectors to project the vector $\mathbf{q}$ onto P, and then form the difference $\mathbf{q} - \mathbf{proj}_P\mathbf{q}$ to obtain $\mathbf{q}_{\perp P}$. A simpler method here is to project $\mathbf{q}$ onto a vector that is known to be orthogonal to P. Since the coefficients of x, y, and z in the equation of the plane provide the components of a normal vector to P, $\mathbf{n} = (2, 1, -2)$ is orthogonal to P. Now, since

$$\text{comp}_{\mathbf{n}}\, \mathbf{q} = \frac{\mathbf{q} \cdot \mathbf{n}}{\|\mathbf{n}\|} = \frac{(3)(2) + (2)(1) + (1)(-2)}{\sqrt{2^2 + 1^2 + (-2)^2}} = 2$$

the distance between P and the point $\mathbf{q}$ is 2. ■

The Gram-Schmidt orthogonalization algorithm. The advantage of an orthonormal basis is clear. As Example 38 showed, the components of a vector relative to an orthonormal basis are very easy to determine: A simple dot product calculation is all that is required. The question is, how do you obtain such a basis? In particular, if B is a basis for a vector space V, how can you transform B into an *orthonormal* basis for V? The process of projecting a vector $\mathbf{v}$ onto a subspace S—then forming the difference $\mathbf{v} - \mathbf{proj}_S \mathbf{v}$ to obtain a vector, $\mathbf{v}_{\perp S}$, orthogonal to S—is the key to the algorithm.

Example 44: Transform the basis $B = \{ \mathbf{v}_1 = (4,\ 2),\ \mathbf{v}_2 = (1,\ 2) \}$ for $\mathbf{R}^2$ into an orthonormal one.

The first step is to keep $\mathbf{v}_1$; it will be normalized later. The second step is to project $\mathbf{v}_2$ onto the subspace spanned by $\mathbf{v}_1$ and then form the difference $\mathbf{v}_2 - \mathbf{proj}_{\mathbf{v}_1} \mathbf{v}_2 = \mathbf{v}_{\perp 1}$. Since

$$\mathbf{proj}_{\mathbf{v}_1} \mathbf{v}_2 = \frac{\mathbf{v}_2 \cdot \mathbf{v}_1}{\mathbf{v}_1 \cdot \mathbf{v}_1} \mathbf{v}_1 = \frac{(1)(4)+(2)(2)}{(4)(4)+(2)(2)}(4,\ 2) = \left(\tfrac{8}{5},\ \tfrac{4}{5} \right)$$

the vector component of $\mathbf{v}_2$ orthogonal to $\mathbf{v}_1$ is

$$\mathbf{v}_{\perp 1} = \mathbf{v}_2 - \mathbf{proj}_{\mathbf{v}_1} \mathbf{v}_2 = (1,\ 2) - \left(\tfrac{8}{5},\ \tfrac{4}{5} \right) = \left(-\tfrac{3}{5},\ \tfrac{6}{5} \right)$$

as illustrated in Figure 50.

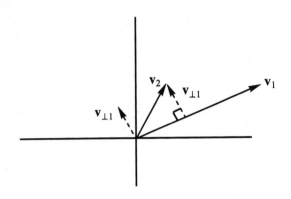

■ Figure 50 ■

The vectors $\mathbf{v}_1$ and $\mathbf{v}_{\perp 1}$ are now normalized:

$$\hat{\mathbf{v}}_1 = \frac{\mathbf{v}_1}{\|\mathbf{v}_1\|} = \frac{(4,\ 2)}{\sqrt{20}} = \left(\tfrac{2}{\sqrt{5}},\ \tfrac{1}{\sqrt{5}}\right)$$

$$\hat{\mathbf{v}}_{\perp 1} = \frac{\mathbf{v}_{\perp 1}}{\|\mathbf{v}_{\perp 1}\|} = \frac{\left(-\tfrac{3}{5},\ \tfrac{6}{5}\right)}{\tfrac{3}{5}\sqrt{5}} = \left(-\tfrac{1}{\sqrt{5}},\ \tfrac{2}{\sqrt{5}}\right)$$

Thus, the basis $B = \{\mathbf{v}_1 = (4,\ 2),\ \mathbf{v}_2 = (1,\ 2)\}$ is transformed into the *orthonormal* basis

$$B' = \left\{\hat{\mathbf{v}}_1 = \left(\tfrac{2}{\sqrt{5}},\ \tfrac{1}{\sqrt{5}}\right),\ \hat{\mathbf{v}}_{\perp 1} = \left(-\tfrac{1}{\sqrt{5}},\ \tfrac{2}{\sqrt{5}}\right)\right\}$$

shown in Figure 51.

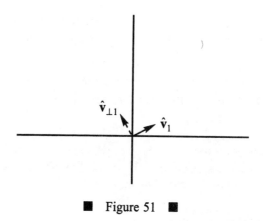

■ Figure 51 ■

The preceding example illustrates the **Gram-Schmidt orthogonalization algorithm** for a basis B consisting of two vectors. It is important to understand that this process not only produces an orthogonal basis B' for the space, but *also preserves the subspaces.* That is, the subspace spanned by the first vector in B' is the same as the subspace spanned by the first vector in B, and the space spanned by the two vectors in B' is the same as the subspace spanned by the two vectors in B.

In general, the Gram-Schmidt orthogonalization algorithm, which transforms a basis, $B = \{v_1, v_2, \ldots, v_r\}$, for a vector space V into an orthogonal basis, $B' = \{w_1, w_2, \ldots, w_r\}$, for V—while preserving the subspaces along the way—proceeds as follows:

Step 1. Set w_1 equal to v_1

Step 2. Project v_2 onto S_1, the space spanned by w_1; then, form the difference $v_2 - \mathbf{proj}_{S_1} v_2$ This is w_2.

Step 3. Project $\mathbf{v}_3$ onto S_2, the space spanned by $\mathbf{w}_1$ and $\mathbf{w}_2$; then, form the difference $\mathbf{v}_3 - \mathbf{proj}_{S_2}\mathbf{v}_3$. This is $\mathbf{w}_3$.

$\vdots$

Step i. Project $\mathbf{v}_i$ onto S_{i-1}, the space spanned by $\mathbf{w}_1, \ldots, \mathbf{w}_{i-1}$; then, form the difference $\mathbf{v}_i - \mathbf{proj}_{S_{i-1}}\mathbf{v}_i$. This is $\mathbf{w}_i$.

$\vdots$

This process continues until Step r, when $\mathbf{w}_r$ is formed, and the orthogonal basis is complete. If an *orthonormal* basis is desired, normalize each of the vectors $\mathbf{w}_i$.

Example 45: Let H be the 3-dimensional subspace of $\mathbf{R}^4$ with basis

$$B = \left\{\mathbf{v}_1 = (0,\ 1,\ -1,\ 0),\ \mathbf{v}_2 = (0,\ 1,\ 0,\ 1),\ \mathbf{v}_3 = (1,\ -1,\ 0,\ 0)\right\}$$

Find an orthogonal basis for H and then—by normalizing these vectors—an orthonormal basis for H. What are the components of the vector $\mathbf{x} = (1,\ 1,\ -1,\ 1)$ relative to this orthonormal basis? What happens if you attempt to find the components of the vector $\mathbf{y} = (1,\ 1,\ 1,\ 1)$ relative to the orthonormal basis?

The first step is to set $\mathbf{w}_1$ equal to $\mathbf{v}_1$. The second step is to project $\mathbf{v}_2$ onto the subspace spanned by $\mathbf{w}_1$ and then form the difference $\mathbf{v}_2 - \mathbf{proj}_{\mathbf{w}_1}\mathbf{v}_2 = \mathbf{w}_2$. Since

$$\mathbf{proj}_{\mathbf{w}_1}\mathbf{v}_2 = \frac{\mathbf{v}_2 \cdot \mathbf{w}_1}{\mathbf{w}_1 \cdot \mathbf{w}_1}\mathbf{w}_1$$

$$= \frac{(0)(0)+(1)(1)+(0)(-1)+(1)(0)}{(0)(0)+(1)(1)+(-1)(-1)+(0)(0)}(0,\ 1,\ -1,\ 0)$$

$$= \left(0,\ \tfrac{1}{2},\ -\tfrac{1}{2},\ 0\right)$$

the vector component of $\mathbf{v}_2$ orthogonal to $\mathbf{w}_1$ is

$$\mathbf{w}_2 = \mathbf{v}_2 - \mathbf{proj}_{\mathbf{w}_1} \mathbf{v}_2$$
$$= (0,\ 1,\ 0,\ 1) - \left(0,\ \tfrac{1}{2},\ -\tfrac{1}{2},\ 0\right)$$
$$= \left(0,\ \tfrac{1}{2},\ \tfrac{1}{2},\ 1\right)$$

Now, for the last step: Project $\mathbf{v}_3$ onto the subspace S_2 spanned by $\mathbf{w}_1$ and $\mathbf{w}_2$ (which is the same as the subspace spanned by $\mathbf{v}_1$ and $\mathbf{v}_2$) and form the difference $\mathbf{v}_3 - \mathbf{proj}_{S_2}\mathbf{v}_3$ to give the vector, $\mathbf{w}_3$, orthogonal to this subspace. Since

$$\mathbf{proj}_{\mathbf{w}_1} \mathbf{v}_3 = \frac{\mathbf{v}_3 \cdot \mathbf{w}_1}{\mathbf{w}_1 \cdot \mathbf{w}_1} \mathbf{w}_1$$

$$= \frac{(1)(0)+(-1)(1)+(0)(-1)+(0)(0)}{(0)(0)+(1)(1)+(-1)(-1)+(0)(0)}(0,\ 1,\ -1,\ 0)$$

$$= \left(0,\ -\tfrac{1}{2},\ \tfrac{1}{2},\ 0\right)$$

and

$$\mathbf{proj}_{\mathbf{w}_2} \mathbf{v}_3 = \frac{\mathbf{v}_3 \cdot \mathbf{w}_2}{\mathbf{w}_2 \cdot \mathbf{w}_2} \mathbf{w}_2$$

$$= \frac{(1)(0)+(-1)(\tfrac{1}{2})+(0)(\tfrac{1}{2})+(0)(1)}{(0)(0)+(\tfrac{1}{2})(\tfrac{1}{2})+(\tfrac{1}{2})(\tfrac{1}{2})+(1)(1)}\left(0,\ \tfrac{1}{2},\ \tfrac{1}{2},\ 1\right)$$

$$= \left(0,\ -\tfrac{1}{6},\ -\tfrac{1}{6},\ -\tfrac{1}{3}\right)$$

and $\{\mathbf{w}_1, \mathbf{w}_2\}$ is an orthogonal basis for S_2, the projection of $\mathbf{v}_3$ onto S_2 is

$$\mathbf{proj}_{S_2} \mathbf{v}_3 = \mathbf{proj}_{\mathbf{w}_1} \mathbf{v}_3 + \mathbf{proj}_{\mathbf{w}_2} \mathbf{v}_3$$

$$= \left(0,\ -\tfrac{1}{2},\ \tfrac{1}{2},\ 0\right) + \left(0,\ -\tfrac{1}{6},\ -\tfrac{1}{6},\ -\tfrac{1}{3}\right)$$

$$= \left(0,\ -\tfrac{2}{3},\ \tfrac{1}{3},\ -\tfrac{1}{3}\right)$$

This gives

$$\mathbf{w}_3 = \mathbf{v}_3 - \mathbf{proj}_{S_2}\mathbf{v}_3$$
$$= (1, \ -1, \ 0, \ 0) - \left(0, \ -\tfrac{2}{3}, \ \tfrac{1}{3}, \ -\tfrac{1}{3}\right)$$
$$= \left(1, \ -\tfrac{1}{3}, \ -\tfrac{1}{3}, \ \tfrac{1}{3}\right)$$

Therefore, the Gram-Schmidt process produces from B the following orthogonal basis for H:

$$B' = \left\{\mathbf{w}_1 = (0, \ 1, \ -1, \ 0), \ \mathbf{w}_2 = \left(0, \ \tfrac{1}{2}, \ \tfrac{1}{2}, \ 1\right), \ \mathbf{w}_3 = \left(1, \ -\tfrac{1}{3}, \ -\tfrac{1}{3}, \ \tfrac{1}{3}\right)\right\}$$

You may verify that these vectors are indeed orthogonal by checking that $\mathbf{w}_1 \cdot \mathbf{w}_2 = \mathbf{w}_1 \cdot \mathbf{w}_3 = \mathbf{w}_2 \cdot \mathbf{w}_3 = 0$ and that the subspaces are preserved along the way:

$$\text{span}\{\mathbf{w}_1\} = \text{span}\{\mathbf{v}_1\}$$
$$\text{span}\{\mathbf{w}_1, \ \mathbf{w}_2\} = \text{span}\{\mathbf{v}_1, \ \mathbf{v}_2\}$$
$$\text{span}\{\mathbf{w}_1, \ \mathbf{w}_2, \ \mathbf{w}_3\} = \text{span}\{\mathbf{v}_1, \ \mathbf{v}_2, \ \mathbf{v}_3\}$$

An orthonormal basis for H is obtained by normalizing the vectors $\mathbf{w}_1$, $\mathbf{w}_2$, and $\mathbf{w}_3$:

$$\hat{\mathbf{w}}_1 = \frac{\mathbf{w}_1}{\|\mathbf{w}_1\|} = \frac{(0, \ 1, \ -1, \ 0)}{\sqrt{2}} = \left(0, \ \tfrac{1}{\sqrt{2}}, \ -\tfrac{1}{\sqrt{2}}, \ 0\right)$$

$$\hat{\mathbf{w}}_2 = \frac{\mathbf{w}_2}{\|\mathbf{w}_2\|} = \frac{\left(0, \ \tfrac{1}{2}, \ \tfrac{1}{2}, \ 1\right)}{\tfrac{1}{2}\sqrt{6}} = \left(0, \ \tfrac{1}{\sqrt{6}}, \ \tfrac{1}{\sqrt{6}}, \ \tfrac{2}{\sqrt{6}}\right)$$

$$\hat{\mathbf{w}}_3 = \frac{\mathbf{w}_3}{\|\mathbf{w}_3\|} = \frac{\left(1, \ -\tfrac{1}{3}, \ -\tfrac{1}{3}, \ \tfrac{1}{3}\right)}{\tfrac{1}{3}\sqrt{12}} = \left(\tfrac{3}{2\sqrt{3}}, \ -\tfrac{1}{2\sqrt{3}}, \ -\tfrac{1}{2\sqrt{3}}, \ \tfrac{1}{2\sqrt{3}}\right)$$

Relative to the orthonormal basis $B'' = \{\hat{\mathbf{w}}_1, \hat{\mathbf{w}}_2, \hat{\mathbf{w}}_3\}$, the vector $\mathbf{x} = (1, 1, -1, 1)$ has components

$$\mathbf{x} \cdot \hat{\mathbf{w}}_1 = (1, 1, -1, 1) \cdot \left(0, \tfrac{1}{\sqrt{2}}, -\tfrac{1}{\sqrt{2}}, 0\right) = \sqrt{2}$$

$$\mathbf{x} \cdot \hat{\mathbf{w}}_2 = (1, 1, -1, 1) \cdot \left(0, \tfrac{1}{\sqrt{6}}, \tfrac{1}{\sqrt{6}}, \tfrac{2}{\sqrt{6}}\right) = \tfrac{2}{\sqrt{6}}$$

$$\mathbf{x} \cdot \hat{\mathbf{w}}_3 = (1, 1, -1, 1) \cdot \left(\tfrac{3}{2\sqrt{3}}, -\tfrac{1}{2\sqrt{3}}, -\tfrac{1}{2\sqrt{3}}, \tfrac{1}{2\sqrt{3}}\right) = \tfrac{2}{\sqrt{3}}$$

These calculations imply that

$$\mathbf{x} = \sqrt{2}\,\hat{\mathbf{w}}_1 + \tfrac{2}{\sqrt{6}}\hat{\mathbf{w}}_2 + \tfrac{2}{\sqrt{3}}\hat{\mathbf{w}}_3$$

a result that is easily verified.

If the components of $\mathbf{y} = (1, 1, 1, 1)$ relative to this basis are desired, you might proceed exactly as above, finding

$$\mathbf{y} \cdot \hat{\mathbf{w}}_1 = (1, 1, 1, 1) \cdot \left(0, \tfrac{1}{\sqrt{2}}, -\tfrac{1}{\sqrt{2}}, 0\right) = 0$$

$$\mathbf{y} \cdot \hat{\mathbf{w}}_2 = (1, 1, 1, 1) \cdot \left(0, \tfrac{1}{\sqrt{6}}, \tfrac{1}{\sqrt{6}}, \tfrac{2}{\sqrt{6}}\right) = \tfrac{4}{\sqrt{6}}$$

$$\mathbf{y} \cdot \hat{\mathbf{w}}_3 = (1, 1, 1, 1) \cdot \left(\tfrac{3}{2\sqrt{3}}, -\tfrac{1}{2\sqrt{3}}, -\tfrac{1}{2\sqrt{3}}, \tfrac{1}{2\sqrt{3}}\right) = \tfrac{1}{\sqrt{3}}$$

These calculations seem to imply that

$$\mathbf{y} = 0\,\hat{\mathbf{w}}_1 + \tfrac{4}{\sqrt{6}}\hat{\mathbf{w}}_2 + \tfrac{1}{\sqrt{3}}\hat{\mathbf{w}}_3 = \tfrac{4}{\sqrt{6}}\hat{\mathbf{w}}_2 + \tfrac{1}{\sqrt{3}}\hat{\mathbf{w}}_3$$

The problem, however, is that this equation is not true, as the following calculation shows:

$$\frac{4}{\sqrt{6}}\,\hat{\mathbf{w}}_2 + \frac{1}{\sqrt{3}}\,\hat{\mathbf{w}}_3 = \frac{4}{\sqrt{6}}\left(0,\ \frac{1}{\sqrt{6}},\ \frac{1}{\sqrt{6}},\ \frac{2}{\sqrt{6}}\right)$$

$$+ \frac{1}{\sqrt{3}}\left(\frac{3}{2\sqrt{3}},\ -\frac{1}{2\sqrt{3}},\ -\frac{1}{2\sqrt{3}},\ \frac{1}{2\sqrt{3}}\right)$$

$$= \left(0,\ \tfrac{2}{3},\ \tfrac{2}{3},\ \tfrac{4}{3}\right) + \left(\tfrac{1}{2},\ -\tfrac{1}{6},\ -\tfrac{1}{6},\ \tfrac{1}{6}\right)$$

$$= \left(\tfrac{1}{2},\ \tfrac{1}{2},\ \tfrac{1}{2},\ \tfrac{3}{2}\right)$$

$$\neq \mathbf{y}$$

What went wrong? The problem is that the vector **y** is not in *H*, so no linear combination of the vectors in any basis for *H* can give **y**. The linear combination

$$\frac{4}{\sqrt{6}}\,\hat{\mathbf{w}}_2 + \frac{1}{\sqrt{3}}\,\hat{\mathbf{w}}_3$$

gives only the projection of **y** onto *H*. ∎

Example 46: If the rows of a matrix form an orthonormal basis for $\mathbf{R}^n$, then the matrix is said to be **orthogonal**. (The term *orthonormal* would have been better, but the terminology is now too well established.) If *A* is an orthogonal matrix, show that $A^{-1} = A^{\mathrm{T}}$.

Let $B = \{\hat{\mathbf{v}}_1,\ \hat{\mathbf{v}}_2,\ \ldots,\ \hat{\mathbf{v}}_n\}$ be an orthonormal basis for $\mathbf{R}^n$ and consider the matrix *A* whose rows are these basis vectors:

$$A = \begin{bmatrix} - & \hat{\mathbf{v}}_1 & \to \\ - & \hat{\mathbf{v}}_2 & \to \\ & \vdots & \\ - & \hat{\mathbf{v}}_n & \to \end{bmatrix}$$

The matrix A^{T} has these basis vectors as its columns:

$$A^{T} = \begin{bmatrix} | & | & & | \\ \hat{\mathbf{v}}_1 & \hat{\mathbf{v}}_2 & \cdots & \hat{\mathbf{v}}_n \\ \downarrow & \downarrow & & \downarrow \end{bmatrix}$$

Since the vectors $\hat{\mathbf{v}}_1, \hat{\mathbf{v}}_2, \ldots, \hat{\mathbf{v}}_n$ are orthonormal,

$$\hat{\mathbf{v}}_i \cdot \hat{\mathbf{v}}_j = \delta_{ij} = \begin{cases} 0 & \text{if } i \neq j \\ 1 & \text{if } i = j \end{cases}$$

Now, because the (i, j) entry of the product AA^{T} is the dot product of row i in A and column j in A^{T},

$$AA^{T} = [\hat{\mathbf{v}}_i \cdot \hat{\mathbf{v}}_j] = [\delta_{ij}] = I$$

Thus, $A^{-1} = A^{T}$. [In fact, the statement $A^{-1} = A^{T}$ is sometimes taken as the definition of an orthogonal matrix (from which it is then shown that the rows of A form an orthonormal basis for $\mathbf{R}^n$).]

An additional fact now follows easily. Assume that A is orthogonal, so $A^{-1} = A^{T}$. Taking the inverse of both sides of this equation gives

$$(A^{-1})^{-1} = (A^{T})^{-1} \quad \Rightarrow \quad A = (A^{T})^{-1} \quad \Rightarrow \quad (A^{T})^{T} = (A^{T})^{-1}$$

which implies that A^{T} is orthogonal (because its transpose equals its inverse). The conclusion

$$A \text{ orthogonal} \quad \Rightarrow \quad A^{T} \text{ orthogonal}$$

means that *if the rows of a matrix form an orthonormal basis for $\mathbf{R}^n$, then so do the columns.* ∎

The Row Space and Column Space of a Matrix

Let A be an m by n matrix. The space spanned by the rows of A is called the **row space** of A, denoted $RS(A)$; it is a subspace of $\mathbf{R}^n$. The space spanned by the columns of A is called the **column space** of A, denoted $CS(A)$; it is a subspace of $\mathbf{R}^m$.

The collection $\{\mathbf{r}_1, \mathbf{r}_2, \ldots, \mathbf{r}_m\}$ consisting of the rows of A may not form a basis for $RS(A)$, because the collection may not be linearly independent. However, a maximal linearly independent subset of $\{\mathbf{r}_1, \mathbf{r}_2, \ldots, \mathbf{r}_m\}$ *does* give a basis for the row space. Since the maximum number of linearly independent rows of A is equal to the rank of A,

$$\dim RS(A) = \operatorname{rank} A \quad (*)$$

Similarly, if $\mathbf{c}_1, \mathbf{c}_2, \ldots, \mathbf{c}_n$ denote the columns of A, then a maximal linearly independent subset of $\{\mathbf{c}_1, \mathbf{c}_2, \ldots, \mathbf{c}_n\}$ gives a basis for the column space of A. But the maximum number of linearly independent columns is also equal to the rank of the matrix, so

$$\dim CS(A) = \operatorname{rank} A \quad (**)$$

Therefore, although $RS(A)$ is a subspace of $\mathbf{R}^n$ and $CS(A)$ is a subspace of $\mathbf{R}^m$, equations $(*)$ and $(**)$ imply that

$$\dim RS(A) = \dim CS(A)$$

even if $m \neq n$.

Example 47: Determine the dimension of, and a basis for, the row space of the matrix

$$B = \begin{bmatrix} 2 & -1 & 3 \\ 1 & 0 & 1 \\ 0 & 2 & -1 \\ 1 & 1 & 4 \end{bmatrix}$$

In Example 23 above, a sequence of elementary row operations reduced this matrix to the echelon matrix

$$\begin{bmatrix} 1 & 0 & 1 \\ 0 & 1 & -1 \\ 0 & 0 & 1 \\ 0 & 0 & 0 \end{bmatrix}$$

The rank of B is 3, so dim $RS(B) = 3$. A basis for $RS(B)$ consists of the nonzero rows in the reduced matrix:

$$\{(1,\ 0,\ 1),\ (0,\ 1,-1),\ (0,\ 0,\ 1)\}$$

Another basis for $RS(B)$, one consisting of some of the original rows of B, is

$$\{\mathbf{r}_1,\ \mathbf{r}_2,\ \mathbf{r}_3\} = \{(2,\ -1,\ 3),\ (1,\ 0,\ 1),\ (0,\ 2,\ -1)\}$$

Note that since the row space is a 3-dimensional subspace of $\mathbf{R}^3$, it must be all of $\mathbf{R}^3$. ■

Criteria for membership in the column space. If A is an $m \times n$ matrix and $\mathbf{x}$ is an n-vector, written as a column matrix, then the product $A\mathbf{x}$ is equal to a linear combination of the columns of A:

$$Ax = \begin{bmatrix} | & | & & | \\ \mathbf{c}_1 & \mathbf{c}_2 & \cdots & \mathbf{c}_n \\ | & | & & | \end{bmatrix} \begin{bmatrix} x_1 \\ x_2 \\ \vdots \\ x_n \end{bmatrix} = x_1\mathbf{c}_1 + x_2\mathbf{c}_2 + \cdots + x_n\mathbf{c}_n \quad (*)$$

By definition, a vector $\mathbf{b}$ in $\mathbf{R}^m$ is in the column space of A if it can be written as a linear combination of the columns of A. That is, $\mathbf{b} \in CS(A)$ precisely when there exist scalars $x_1, x_2, \ldots, x_n$ such that

$$x_1\mathbf{c}_1 + x_2\mathbf{c}_2 + \cdots + x_n\mathbf{c}_n = \mathbf{b} \quad (**)$$

Combining (*) and (**), then, leads to the following conclusion:

$$\boxed{\mathbf{b} \in CS(A) \quad \Leftrightarrow \quad A\mathbf{x} = \mathbf{b} \text{ is consistent}}$$

Example 48: For what value of b is the vector $\mathbf{b} = (1, 2, 3, b)^\mathsf{T}$ in the column space of the following matrix?

$$A = \begin{bmatrix} 2 & 3 & 3 \\ 0 & -4 & -5 \\ 6 & 3 & 0 \\ 1 & 1 & 3 \end{bmatrix}$$

Form the augmented matrix $[A|\mathbf{b}]$ and reduce:

$$[A|\mathbf{b}] = \begin{bmatrix} 2 & 3 & 3 & 1 \\ 0 & -4 & -5 & 2 \\ 6 & 3 & 0 & 3 \\ 1 & 1 & 3 & b \end{bmatrix}$$

$$\xrightarrow{\mathbf{r}_1 \leftrightarrow \mathbf{r}_4} \begin{bmatrix} 1 & 1 & 3 & b \\ 0 & -4 & -5 & 2 \\ 6 & 3 & 0 & 3 \\ 2 & 3 & 3 & 1 \end{bmatrix}$$

$$\xrightarrow[\substack{-6\mathbf{r}_1 \text{ added to } \mathbf{r}_3 \\ -2\mathbf{r}_1 \text{ added to } \mathbf{r}_4}]{} \begin{bmatrix} 1 & 1 & 3 & b \\ 0 & -4 & -5 & 2 \\ 0 & -3 & -18 & 3-6b \\ 0 & 1 & -3 & 1-2b \end{bmatrix}$$

$$\xrightarrow{\mathbf{r}_2 \leftrightarrow \mathbf{r}_4} \begin{bmatrix} 1 & 1 & 3 & b \\ 0 & 1 & -3 & 1-2b \\ 0 & -3 & -18 & 3-6b \\ 0 & -4 & -5 & 2 \end{bmatrix}$$

$$\xrightarrow[\substack{3\mathbf{r}_2 \text{ added to } \mathbf{r}_3 \\ 4\mathbf{r}_2 \text{ added to } \mathbf{r}_4}]{} \begin{bmatrix} 1 & 1 & 3 & b \\ 0 & 1 & -3 & 1-2b \\ 0 & 0 & -27 & 6-12b \\ 0 & 0 & -17 & 6-8b \end{bmatrix}$$

$$\xrightarrow{-\frac{17}{27}\mathbf{r}_3 \text{ added to } \mathbf{r}_4} \begin{bmatrix} 1 & 1 & 3 & b \\ 0 & 1 & -3 & 1-2b \\ 0 & 0 & -27 & 6-12b \\ 0 & 0 & 0 & (6-8b)-\frac{17}{27}(6-12b) \end{bmatrix} = [A'|\mathbf{b}']$$

Because of the bottom row of zeros in A' (the reduced form of A), the bottom entry in the last column must also be 0—giving a complete row of zeros at the bottom of $[A'|\mathbf{b}']$—in order for the system $A\mathbf{x} = \mathbf{b}$ to have a solution. Setting $(6-8b)-(17/27)(6-12b)$ equal to 0 and solving for b yields

$$(6-8b)-\tfrac{17}{27}(6-12b)=0$$
$$27(6-8b)=17(6-12b)$$
$$162-216b=102-204b$$
$$-12b=-60$$
$$b=5$$

Therefore, $\mathbf{b} = (1, 2, 3, b)^{\mathrm{T}}$ is in $CS(A)$ if and only if $b = 5$. ∎

Since elementary row operations do not change the rank of a matrix, it is clear that in the calculation above, rank A = rank A' and rank $[A|\mathbf{b}]$ = rank $[A'|\mathbf{b}']$. (Since the bottom row of A' consisted entirely of zeros, rank A' = 3, implying rank A = 3 also.) With b = 5, the bottom row of $[A'|\mathbf{b}']$ also consists entirely of zeros, giving rank $[A'|\mathbf{b}']$ = 3. However, if b were not equal to 5, then the bottom row of $[A'|\mathbf{b}']$ would not consist entirely of zeros, and the rank of $[A'|\mathbf{b}']$ would have been 4, not 3. This example illustrates the following general fact: When $\mathbf{b}$ is in $CS(A)$, the rank of $[A|\mathbf{b}]$ is the same as the rank of A; and, conversely, when $\mathbf{b}$ is not in $CS(A)$, the rank of $[A|\mathbf{b}]$ is not the same as (it's strictly greater than) the rank of A. Therefore, an equivalent criterion for membership in the column space of a matrix reads as follows:

$$\mathbf{b} \in CS(A) \quad \Leftrightarrow \quad \text{rank } A = \text{rank}\left[A|\mathbf{b}\right]$$

Example 49: Determine the dimension of, and a basis for, the column space of the matrix

$$B = \begin{bmatrix} 2 & -1 & 3 \\ 1 & 0 & 1 \\ 0 & 2 & -1 \\ 1 & 1 & 4 \end{bmatrix}$$

from Example 47 above.

Because the dimension of the column space of a matrix always equals the dimension of its row space, $CS(B)$ must also have dimension 3: $CS(B)$ is a 3-dimensional subspace of $\mathbf{R}^4$. Since B contains only 3 columns, these columns must be linearly independent and therefore form a basis:

$$\text{basis for } CS(B) = \left\{ \begin{bmatrix} 2 \\ 1 \\ 0 \\ 1 \end{bmatrix}, \begin{bmatrix} -1 \\ 0 \\ 2 \\ 1 \end{bmatrix}, \begin{bmatrix} 3 \\ 1 \\ -1 \\ 4 \end{bmatrix} \right\}$$ ∎

Example 50: Find a basis for the column space of the matrix

$$A = \begin{bmatrix} 1 & 2 & -1 & 3 & 1 \\ 2 & 0 & -6 & 2 & -2 \\ -3 & 1 & 10 & -2 & 4 \end{bmatrix}$$

Since the column space of A consists precisely of those vectors $\mathbf{b}$ such that $A\mathbf{x} = \mathbf{b}$ is a solvable system, one way to determine a basis for $CS(A)$ would be to first find the space of all vectors $\mathbf{b}$ such that $A\mathbf{x} = \mathbf{b}$ is consistent, then constructing a basis for this space. However, an elementary observation suggests a simpler approach: *Since the columns of A are the rows of A^T, finding a basis for CS(A) is equivalent to finding a basis for RS(A^T)*. Row-reducing A^T yields

$$A^T = \begin{bmatrix} 1 & 2 & -3 \\ 2 & 0 & 1 \\ -1 & -6 & 10 \\ 3 & 2 & -2 \\ 1 & -2 & 4 \end{bmatrix} \xrightarrow[\substack{-3r_1 \text{ added to } r_4 \\ -r_1 \text{ added to } r_5}]{\substack{-2r_1 \text{ added to } r_2 \\ r_1 \text{ added to } r_3}} \begin{bmatrix} 1 & 2 & -3 \\ 0 & -4 & 7 \\ 0 & -4 & 7 \\ 0 & -4 & 7 \\ 0 & -4 & 7 \end{bmatrix}$$

$$\xrightarrow[\substack{-r_2 \text{ added to } r_5}]{\substack{-r_2 \text{ added to } r_3 \\ -r_2 \text{ added to } r_4}} \begin{bmatrix} 1 & 2 & -3 \\ 0 & -4 & 7 \\ 0 & 0 & 0 \\ 0 & 0 & 0 \\ 0 & 0 & 0 \end{bmatrix}$$

Since there are two nonzero rows left in the reduced form of A^T, the rank of A^T is 2, so

$$\dim RS(A^T) = 2 = \dim CS(A)$$

Furthermore, since $\{\mathbf{v}_1, \mathbf{v}_2\} = \{(1, 2, -3), (0, -4, 7)\}$ is a basis for $RS(A^T)$, the collection

$$\{\mathbf{v}_1^T, \mathbf{v}_2^T\} = \left\{ \begin{bmatrix} 1 \\ 2 \\ -3 \end{bmatrix}, \begin{bmatrix} 0 \\ -4 \\ 7 \end{bmatrix} \right\}$$

is a basis for $CS(A)$, a 2-dimensional subspace of $\mathbf{R}^3$. ∎

The Rank Plus Nullity Theorem

Let A be a matrix. Recall that the dimension of its column space (and row space) is called the rank of A. The dimension of its nullspace is called the **nullity** of A. The connection between these dimensions is illustrated in the following example.

Example 51: Find the nullspace of the matrix

$$A = \begin{bmatrix} 0 & 1 & 2 & -1 & 3 \\ 1 & -1 & -3 & 1 & 1 \\ 4 & 0 & 0 & 1 & -2 \\ 2 & 3 & 8 & -2 & -1 \end{bmatrix}$$

The nullspace of A is the solution set of the homogeneous equation $A\mathbf{x} = \mathbf{0}$. To solve this equation, the following elementary row operations are performed to reduce A to echelon form:

$$A = \begin{bmatrix} 0 & 1 & 2 & -1 & 3 \\ 1 & -1 & -3 & 1 & 1 \\ 4 & 0 & 0 & 1 & -2 \\ 2 & 3 & 8 & -2 & -1 \end{bmatrix} \xrightarrow{\mathbf{r}_1 \leftrightarrow \mathbf{r}_2} \begin{bmatrix} 1 & -1 & -3 & 1 & 1 \\ 0 & 1 & 2 & -1 & 3 \\ 4 & 0 & 0 & 1 & -2 \\ 2 & 3 & 8 & -2 & -1 \end{bmatrix}$$

$$\xrightarrow[-2\mathbf{r}_1 \text{ added to } \mathbf{r}_4]{-4\mathbf{r}_1 \text{ added to } \mathbf{r}_3} \begin{bmatrix} 1 & -1 & -3 & 1 & 1 \\ 0 & 1 & 2 & -1 & 3 \\ 0 & 4 & 12 & -3 & -6 \\ 0 & 5 & 14 & -4 & -3 \end{bmatrix}$$

$$\xrightarrow[\substack{-4\mathbf{r}_2 \text{ added to } \mathbf{r}_3 \\ -5\mathbf{r}_2 \text{ added to } \mathbf{r}_4}]{} \begin{bmatrix} 1 & -1 & -3 & 1 & 1 \\ 0 & 1 & 2 & -1 & 3 \\ 0 & 0 & 4 & 1 & -18 \\ 0 & 0 & 4 & 1 & -18 \end{bmatrix}$$

$$\xrightarrow[\substack{-\mathbf{r}_3 \text{ added to } \mathbf{r}_4}]{} \begin{bmatrix} 1 & -1 & -2 & 1 & 1 \\ 0 & 1 & 2 & -1 & 3 \\ 0 & 0 & 4 & 1 & -18 \\ 0 & 0 & 0 & 0 & 0 \end{bmatrix} = A'$$

Therefore, the solution set of $A\mathbf{x} = \mathbf{0}$ is the same as the solution set of $A'\mathbf{x} = \mathbf{0}$:

$$\begin{bmatrix} 1 & -1 & -2 & 1 & 1 \\ 0 & 1 & 2 & -1 & 3 \\ 0 & 0 & 4 & 1 & -18 \\ 0 & 0 & 0 & 0 & 0 \end{bmatrix} \begin{bmatrix} x_1 \\ x_2 \\ x_3 \\ x_4 \\ x_5 \end{bmatrix} = \begin{bmatrix} 0 \\ 0 \\ 0 \\ 0 \\ 0 \end{bmatrix}$$

With only three nonzero rows in the coefficient matrix, there are really only three constraints on the variables, leaving $5 - 3 = 2$ of the variables free. Let x_4 and x_5 be the free variables. Then the third row of A' implies

$$4x_3 + x_4 - 18x_5 = 0 \implies x_3 = -\tfrac{1}{4}x_4 + \tfrac{9}{2}x_5$$

The second row now yields

$$x_2 + 2(-\tfrac{1}{4}x_4 + \tfrac{9}{2}x_5) - x_4 + 3x_5 = 0$$
$$2x_2 - x_4 + 18x_5 - 2x_4 + 6x_5 = 0$$
$$2x_2 - 3x_4 + 24x_5 = 0$$
$$x_2 = \tfrac{3}{2}x_4 - 12x_5$$

from which the first row gives

$$x_1 - (\tfrac{3}{2}x_4 - 12x_5) - 2(-\tfrac{1}{4}x_4 + \tfrac{9}{2}x_5) + x_4 + x_5 = 0$$
$$2x_1 - 3x_4 + 24x_5 + x_4 - 18x_5 + 2x_4 + 2x_5 = 0$$
$$2x_1 + 8x_5 = 0$$
$$x_1 = -4x_5$$

Therefore, the solutions of the equation $A\mathbf{x} = \mathbf{0}$ are those vectors of the form

$$(x_1, x_2, x_3, x_4, x_5)^{\mathrm{T}} = (-4x_5, \tfrac{3}{2}x_4 - 12x_5, -\tfrac{1}{4}x_4 + \tfrac{9}{2}x_5, x_4, x_5)^{\mathrm{T}}$$

To clear this expression of fractions, let $t_1 = \tfrac{1}{4}x_4$ and $t_2 = \tfrac{1}{2}x_5$; then, those vectors $\mathbf{x}$ in $\mathbf{R}^5$ that satisfy the homogeneous system $A\mathbf{x} = \mathbf{0}$ have the form

$$\mathbf{x} = (-8t_2, 6t_1 - 24t_2, -t_1 + 9t_2, 4t_1, 2t_2)^{\mathrm{T}}$$
$$= t_1(0, 6, -1, 4, 0)^{\mathrm{T}} + t_2(-8, -24, 9, 0, 2)^{\mathrm{T}}$$

Note in particular that the number of free variables—the number of parameters in the general solution—is the dimension of the nullspace (which is 2 in this case). Also, the rank of this matrix, which is the number of nonzero rows in its echelon form, is 3. The sum of the nullity and the rank, 2 + 3, is equal to the number of columns of the matrix. ∎

The connection between the rank and nullity of a matrix, illustrated in the preceding example, actually holds for *any* matrix:

The Rank Plus Nullity Theorem. Let A be an m by n matrix, with rank r and nullity ℓ. Then $r + \ell = n$; that is,

   rank A + nullity A = the number of columns of A

Proof. Consider the matrix equation $A\mathbf{x} = \mathbf{0}$ and assume that A has been reduced to echelon form, A'. First, note that the elementary row operations which reduce A to A' do not change the row space or, consequently, the rank of A. Second, it is clear that the number of components in $\mathbf{x}$ is n, the number of columns of A and of A'. Since A' has only r nonzero rows (because its rank is r), $n - r$ of the variables $x_1, x_2, \ldots, x_n$ in $\mathbf{x}$ are free. But the number of free variables—that is, the number of parameters in the general solution of $A\mathbf{x} = \mathbf{0}$—is the nullity of A. Thus, nullity $A = n - r$, and the statement of the theorem, $r + \ell = r + (n - r) = n$, follows immediately. ■

Example 52: If A is a 5×6 matrix with rank 2, what is the dimension of the nullspace of A?

Since the nullity is the difference between the number of columns of A and the rank of A, the nullity of this matrix is $6 - 2 = 4$. Its nullspace is a 4-dimensional subspace of $\mathbf{R}^6$. ■

Example 53: Find a basis for the nullspace of the matrix

$$A = \begin{bmatrix} 1 & -2 & 0 & 4 \\ 3 & 1 & 1 & 0 \\ -1 & -5 & -1 & 8 \end{bmatrix}$$

Recall that for a given m by n matrix A, the set of all solutions of the homogeneous system $A\mathbf{x} = \mathbf{0}$ forms a subspace of $\mathbf{R}^n$ called the nullspace of A. To solve $A\mathbf{x} = \mathbf{0}$, the matrix A is row reduced:

$$A = \begin{bmatrix} 1 & -2 & 0 & 4 \\ 3 & 1 & 1 & 0 \\ -1 & -5 & -1 & 8 \end{bmatrix} \xrightarrow[\;r_1 \text{ added to } r_3\;]{-3r_1 \text{ added to } r_2} \begin{bmatrix} 1 & -2 & 0 & 4 \\ 0 & 7 & 1 & -12 \\ 0 & -7 & -1 & 12 \end{bmatrix}$$

$$\xrightarrow{\;r_2 \text{ added to } r_3\;} \begin{bmatrix} 1 & -2 & 0 & 4 \\ 0 & 7 & 1 & -12 \\ 0 & 0 & 0 & 0 \end{bmatrix}$$

Clearly, the rank of A is 2. Since A has 4 columns, the rank plus nullity theorem implies that the nullity of A is $4 - 2 = 2$. Let x_3 and x_4 be the free variables. The second row of the reduced matrix gives

$$7x_2 + x_3 - 12x_4 = 0 \quad \Rightarrow \quad x_2 = \tfrac{1}{7}(-x_3 + 12x_4)$$

and the first row then yields

$$x_1 - 2\left[\tfrac{1}{7}(-x_3 + 12x_4)\right] + 4x_4 = 0 \quad \Rightarrow \quad x_1 = -\tfrac{1}{7}(2x_3 + 4x_4)$$

Therefore, the vectors $\mathbf{x}$ in the nullspace of A are precisely those of the form

$$\mathbf{x} = \begin{bmatrix} -\frac{1}{7}(2x_3 + 4x_4) \\ \frac{1}{7}(-x_3 + 12x_4) \\ x_3 \\ x_4 \end{bmatrix}$$

which can be expressed as follows:

$$\mathbf{x} = \begin{bmatrix} -\frac{1}{7}(2x_3) \\ \frac{1}{7}(-x_3) \\ x_3 \\ 0 \end{bmatrix} + \begin{bmatrix} -\frac{1}{7}(4x_4) \\ \frac{1}{7}(12x_4) \\ 0 \\ x_4 \end{bmatrix} = \frac{1}{7}x_3 \begin{bmatrix} -2 \\ -1 \\ 7 \\ 0 \end{bmatrix} + \frac{1}{7}x_4 \begin{bmatrix} -4 \\ 12 \\ 0 \\ 7 \end{bmatrix}$$

If $t_1 = \frac{1}{7}x_3$ and $t_2 = \frac{1}{7}x_4$, then $\mathbf{x} = t_1(-2, -1, 7, 0)^T + t_2(-4, 12, 0, 7)^T$, so

$$N(A) = \operatorname{span}\left\{ \begin{bmatrix} -2 \\ -1 \\ 7 \\ 0 \end{bmatrix}, \begin{bmatrix} -4 \\ 12 \\ 0 \\ 7 \end{bmatrix} \right\}$$

Since the two vectors in this collection are linearly independent (because neither is a multiple of the other), they form a basis for $N(A)$:

$$\text{basis for } N(A) = \left\{ \begin{bmatrix} -2 \\ -1 \\ 7 \\ 0 \end{bmatrix}, \begin{bmatrix} -4 \\ 12 \\ 0 \\ 7 \end{bmatrix} \right\} \quad \blacksquare$$

Other Real Euclidean Vector Spaces and the Concept of Isomorphism

The idea of a vector space can be extended to include objects that you would not initially consider to be ordinary vectors.

Matrix spaces. Consider the set $M_{2\times3}(\mathbf{R})$ of 2 by 3 matrices with real entries. This set is closed under addition, since the sum of a pair of 2 by 3 matrices is again a 2 by 3 matrix, and when such a matrix is multiplied by a real scalar, the resulting matrix is in the set also. Since $M_{2\times3}(\mathbf{R})$, with the usual algebraic operations, is closed under addition and scalar multiplication, it is a real Euclidean vector space. The objects in the space—the "vectors"—are now matrices.

Since $M_{2\times3}(\mathbf{R})$ is a vector space, what is its dimension? First, note that any 2 by 3 matrix is a unique linear combination of the following six matrices:

$$E_1 = \begin{bmatrix} 1 & 0 & 0 \\ 0 & 0 & 0 \end{bmatrix}, \quad E_2 = \begin{bmatrix} 0 & 1 & 0 \\ 0 & 0 & 0 \end{bmatrix}, \quad E_3 = \begin{bmatrix} 0 & 0 & 1 \\ 0 & 0 & 0 \end{bmatrix},$$

$$E_4 = \begin{bmatrix} 0 & 0 & 0 \\ 1 & 0 & 0 \end{bmatrix}, \quad E_5 = \begin{bmatrix} 0 & 0 & 0 \\ 0 & 1 & 0 \end{bmatrix}, \quad E_6 = \begin{bmatrix} 0 & 0 & 0 \\ 0 & 0 & 1 \end{bmatrix}$$

Therefore, they span $M_{2\times3}(\mathbf{R})$. Furthermore, these "vectors" are linearly independent: none of these matrices is a linear combination of the others. (Alternatively, the only way $k_1 E_1 + k_2 E_2 + k_3 E_3 + k_4 E_4 + k_5 E_5 + k_6 E_6$ will give the 2 by 3 zero matrix is if each scalar coefficient, k_i, in this combination is zero.) These six "vectors" therefore form a basis for $M_{2\times3}(\mathbf{R})$, so dim $M_{2\times3}(\mathbf{R}) = 6$.

If the entries in a given 2 by 3 matrix are written out in a single row (or column), the result is a vector in $\mathbf{R}^6$. For example,

$$\begin{bmatrix} 1 & 3 & -4 \\ -2 & -1 & 0 \end{bmatrix} \xrightarrow{\text{gives}} (1,\ 3,\ -4,\ -2,\ -1,\ 0)$$

The rule here is simple: Given a 2 by 3 matrix, form a 6-vector by writing the entries in the first row of the matrix followed by the entries in the second row. Then, to every matrix in $M_{2\times3}(\mathbf{R})$ there corresponds a unique vector in $\mathbf{R}^6$, and vice versa. This one-to-one correspondence between $M_{2\times3}(\mathbf{R})$ and $\mathbf{R}^6$,

$$\begin{bmatrix} a & b & c \\ d & e & f \end{bmatrix} \underset{\varphi^{-1}}{\overset{\varphi}{\rightleftarrows}} (a,\ b,\ c,\ d,\ e,\ f)$$

is compatible with the vector space operations of addition and scalar multiplication. This means that

$$\varphi(A+B) = \varphi(A) + \varphi(B)$$
$$\varphi(kA) = k\varphi(A)$$

The conclusion is that the spaces $M_{2\times3}(\mathbf{R})$ and $\mathbf{R}^6$ are *structurally identical*, that is, **isomorphic**, a fact which is denoted $M_{2\times3}(\mathbf{R}) \cong \mathbf{R}^6$. One consequence of this structural identity is that under the mapping φ—the *isomorphism*—each basis "vector" E_i given above for $M_{2\times3}(\mathbf{R})$ corresponds to the standard basis vector $\mathbf{e}_i$ for $\mathbf{R}^6$. The only real difference between the spaces $\mathbf{R}^6$ and $M_{2\times3}(\mathbf{R})$ is in the notation: The six entries denoting an element in $\mathbf{R}^6$ are written as a single row (or column), while the six entries denoting an element in $M_{2\times3}(\mathbf{R})$ are written in two rows of three entries each.

This example can be generalized further. If m and n are any positive integers, then the set of real m by n matrices, $M_{m\times n}(\mathbf{R})$, is isomorphic to $\mathbf{R}^{mn}$, which implies that dim $M_{m\times n}(\mathbf{R}) = mn$.

Example 54: Consider the subset $S_{3\times 3}(\mathbf{R}) \subset M_{3\times 3}(\mathbf{R})$ consisting of the symmetric matrices, that is, those which equal their transpose. Show that $S_{3\times 3}(\mathbf{R})$ is actually a subspace of $M_{3\times 3}(\mathbf{R})$ and then determine the dimension and a basis for this subspace. What is the dimension of the subspace $S_{n\times n}(\mathbf{R})$ of symmetric n by n matrices?

Since $M_{3\times 3}(\mathbf{R})$ is a Euclidean vector space (isomorphic to $\mathbf{R}^9$), all that is required to establish that $S_{3\times 3}(\mathbf{R})$ is a subspace is to show that it is closed under addition and scalar multiplication. If $A = A^T$ and $B = B^T$, then $(A + B)^T = A^T + B^T = A + B$, so $A + B$ is symmetric; thus, $S_{3\times 3}(\mathbf{R})$ is closed under addition. Furthermore, if A is symmetric, then $(kA)^T = kA^T = kA$, so kA is symmetric, showing that $S_{3\times 3}(\mathbf{R})$ is also closed under scalar multiplication.

As for the dimension of this subspace, note that the 3 entries on the diagonal (①, ②, and ③ in the diagram below), and the 2 + 1 entries above the diagonal (④, ⑤, and ⑥) can be chosen arbitrarily, but the other 1 + 2 entries below the diagonal are then completely determined by the symmetry of the matrix:

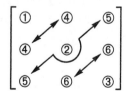

Therefore, there are only $3 + 2 + 1 = 6$ degrees of freedom in the selection of the nine entries in a 3 by 3 symmetric matrix. The conclusion, then, is that dim $S_{3 \times 3}(\mathbf{R}) = 6$. A basis for $S_{3 \times 3}(\mathbf{R})$ consists of the six 3 by 3 matrices

$$\begin{bmatrix} 1 & & \\ & & \\ & & \end{bmatrix}, \quad \begin{bmatrix} & & \\ & 1 & \\ & & \end{bmatrix}, \quad \begin{bmatrix} & & \\ & & \\ & & 1 \end{bmatrix},$$

$$\begin{bmatrix} & 1 & \\ 1 & & \\ & & \end{bmatrix}, \quad \begin{bmatrix} & & 1 \\ & & \\ 1 & & \end{bmatrix}, \quad \begin{bmatrix} & & \\ & & 1 \\ & 1 & \end{bmatrix}$$

In general, there are $n + (n - 1) + \cdots + 2 + 1 = \frac{1}{2}n(n + 1)$ degrees of freedom in the selection of entries in an n by n symmetric matrix, so dim $S_{n \times n}(\mathbf{R}) = \frac{1}{2}n(n + 1)$. ∎

Polynomial spaces. A polynomial of degree n is an expression of the form

$$a_0 + a_1 x + a_2 x^2 + \cdots + a_n x^n$$

where the coefficients a_i are real numbers. The set of all such polynomials of degree $\leq n$ is denoted P_n. With the usual algebraic operations, P_n is a vector space, because it is closed under addition (the sum of any two polynomials of degree $\leq n$ is again a polynomial of degree $\leq n$) and scalar multiplication (a scalar times a polynomial of degree $\leq n$ is still a polynomial of degree $\leq n$). The "vectors" are now polynomials.

There is a simple isomorphism between P_n and $\mathbf{R}^{n+1}$:

$$a_0 + a_1 x + a_2 x^2 + \cdots + a_n x^n \xrightleftharpoons[\varphi^{-1}]{\varphi} (a_0,\, a_1,\, a_2,\, \ldots,\, a_n) \in \mathbf{R}^{n+1}$$

This mapping is clearly a one-to-one correspondence and compatible with the vector space operations. Therefore, $P_n \cong \mathbf{R}^{n+1}$, which immediately implies dim $P_n = n + 1$. The standard basis for P_n, $\{1,\, x,\, x^2,\, \ldots,\, x^n\}$, comes from the standard basis for $\mathbf{R}^{n+1}$, $\{\mathbf{e}_1,\, \mathbf{e}_2,\, \mathbf{e}_3,\, \ldots,\, \mathbf{e}_{n+1}\}$, under the mapping φ^{-1}:

$$1 \xleftarrow{\varphi^{-1}} (1, 0, 0, \ldots, 0) = \mathbf{e}_1$$
$$x \xleftarrow{\varphi^{-1}} (0, 1, 0, \ldots, 0) = \mathbf{e}_2$$
$$x^2 \xleftarrow{\varphi^{-1}} (0, 0, 1, \ldots, 0) = \mathbf{e}_3$$
$$\vdots$$
$$x^n \xleftarrow{\varphi^{-1}} (0, 0, 0, \ldots, 1) = \mathbf{e}_{n+1}$$

Example 55: Are the polynomials $\mathbf{p}_1 = 2 - x$, $\mathbf{p}_2 = 1 + x + x^2$, and $\mathbf{p}_3 = 3x - 2x^2$ from P_2 linearly independent?

One way to answer this question is to recast it in terms of $\mathbf{R}^3$, since P_2 is isomorphic to $\mathbf{R}^3$. Under the isomorphism given above, $\mathbf{p}_1$ corresponds to the vector $\mathbf{v}_1 = (2, -1, 0)$, $\mathbf{p}_2$ corresponds to $\mathbf{v}_2 = (1, 1, 1)$, and $\mathbf{p}_3$ corresponds to $\mathbf{v}_3 = (0, 3, -2)$. Therefore, asking whether the polynomials $\mathbf{p}_1$, $\mathbf{p}_2$, and $\mathbf{p}_3$ are independent in the space P_2 is exactly the same as asking whether the vectors $\mathbf{v}_1$, $\mathbf{v}_2$, and $\mathbf{v}_3$ are independent in the space $\mathbf{R}^3$. Put yet another way, does the matrix

$$\begin{bmatrix} - & \mathbf{v}_1 & \rightarrow \\ - & \mathbf{v}_2 & \rightarrow \\ - & \mathbf{v}_3 & \rightarrow \end{bmatrix} = \begin{bmatrix} 2 & -1 & 0 \\ 1 & 1 & 1 \\ 0 & 3 & -2 \end{bmatrix}$$

have full rank (that is, rank 3)? A few elementary row operations reduce this matrix to an echelon form with three nonzero rows:

$$\begin{bmatrix} 2 & -1 & 0 \\ 1 & 1 & 1 \\ 0 & 3 & -2 \end{bmatrix} \xrightarrow{\ \mathbf{r}_1 \leftrightarrow \mathbf{r}_2\ } \begin{bmatrix} 1 & 1 & 1 \\ 2 & -1 & 0 \\ 0 & 3 & -2 \end{bmatrix}$$

$$\xrightarrow{\ -2\mathbf{r}_1 \text{ added to } \mathbf{r}_2\ } \begin{bmatrix} 1 & 1 & 1 \\ 0 & -3 & -2 \\ 0 & 3 & -2 \end{bmatrix}$$

$$\xrightarrow{\ \mathbf{r}_2 \text{ added to } \mathbf{r}_3\ } \begin{bmatrix} 1 & 1 & 1 \\ 0 & -3 & -2 \\ 0 & 0 & -4 \end{bmatrix}$$

Thus, the vectors—either $\mathbf{v}_1$, $\mathbf{v}_2$, $\mathbf{v}_3$ in $\mathbf{R}^3$ or $\mathbf{p}_1$, $\mathbf{p}_2$, $\mathbf{p}_3$ in P_2— are indeed independent. ■

Function spaces. Let A be a subset of the real line and consider the collection of all real-valued functions f defined on A. This collection of functions is denoted $\mathbf{R}^A$. It is certainly closed under addition (the sum of two such functions is again such a function) and scalar multiplication (a real scalar multiple of a function in this set is also a function in this set), so $\mathbf{R}^A$ is a vector space; the "vectors" are now functions. Unlike each of the matrix and polynomial spaces described above, this vector space has no finite basis (for example, $\mathbf{R}^A$ contains P_n for *every* n); $\mathbf{R}^A$ is infinite-dimensional. The real-valued functions which are continuous on A, or those which are bounded on A, are subspaces of $\mathbf{R}^A$ which are also infinite-dimensional.

Example 56: Are the functions $\mathbf{f}_1 = \sin^2 x$, $\mathbf{f}_2 = \cos^2 x$, and $\mathbf{f}_3 \equiv 3$ linearly independent in the space of continuous functions defined everywhere on the real line?

Does there exist a nontrivial linear combination of $\mathbf{f}_1$, $\mathbf{f}_2$, and $\mathbf{f}_3$ that gives the zero function? Yes: $3\mathbf{f}_1 + 3\mathbf{f}_2 - \mathbf{f}_3 \equiv \mathbf{0}$. This establishes that these three functions are not independent. ∎

Example 57: Let $C^2(\mathbf{R})$ denote the vector space of all real-valued functions defined everywhere on the real line that possess a continuous second derivative. Show that the set of solutions of the differential equation $y'' + y = 0$ is a 2-dimensional subspace of $C^2(\mathbf{R})$.

From the theory of homogeneous differential equations with constant coefficients, it is known that the equation $y'' + y = 0$ is satisfied by $y_1 = \cos x$ and $y_2 = \sin x$ and, more generally, by any linear combination, $y = c_1 \cos x + c_2 \sin x$, of these functions. Since $y_1 = \cos x$ and $y_2 = \sin x$ are linearly independent (neither is a constant multiple of the other) and they span the space S of solutions, a basis for S is $\{\cos x, \sin x\}$, which contains two elements. Thus,

$$\dim S = \dim\left\{y \in C^2(\mathbf{R}): y'' + y = 0\right\} = 2$$

as desired. ∎

Associated with each square matrix A is a number called its **determinant**, denoted by det A or by the symbol $|A|$. The purpose here is to give the definition of the determinant, to illustrate how it is evaluated, and to discuss some of its applications. Throughout this section, *all matrices are square*; the determinant of a nonsquare matrix is meaningless.

Definitions of the Determinant

The determinant function can be defined by essentially two different methods. The advantage of the first definition—one which uses *permutations*—is that it provides an actual formula for det A, a fact of theoretical importance. The disadvantage is that, quite frankly, no one actually computes a determinant by this method.

Method 1 for defining the determinant. If n is a positive integer, then a **permutation** of the set $S = \{1, 2, \ldots, n\}$ is defined to be a bijective function—that is, a one-to-one correspondence—σ, from S to S. For example, let $S = \{1, 2, 3\}$ and define a permutation σ of S as follows:

$$1 \overset{\sigma}{\mapsto} 3, \quad 2 \overset{\sigma}{\mapsto} 1, \quad 3 \overset{\sigma}{\mapsto} 2$$

Since $\sigma(1) = 3$, $\sigma(2) = 1$, and $\sigma(3) = 2$, the permutation σ maps the elements 1, 2, 3 into 3, 1, 2. *Intuitively, then, a permutation of the set $S = \{1, 2, \ldots, n\}$ provides a rearrangement of the numbers 1, 2, . . ., n.* Another permutation, σ', of the set S is defined as follows:

$$1 \overset{\sigma'}{\mapsto} 2, \quad 2 \overset{\sigma'}{\mapsto} 1, \quad 3 \overset{\sigma'}{\mapsto} 3$$

This permutation maps the elements 1, 2, 3 into 2, 1, 3, respectively. This result is written

$$(1, 2, 3) \overset{\sigma'}{\mapsto} (2, 1, 3)$$

Example 1: In all, there are six possible permutations of the 3-element set $S = \{1, 2, 3\}$:

$$(1, 2, 3) \overset{\sigma_1}{\mapsto} (1, 2, 3) \qquad (1, 2, 3) \overset{\sigma_4}{\mapsto} (2, 3, 1)$$

$$(1, 2, 3) \overset{\sigma_2}{\mapsto} (1, 3, 2) \qquad (1, 2, 3) \overset{\sigma_5}{\mapsto} (3, 1, 2)$$

$$(1, 2, 3) \overset{\sigma_3}{\mapsto} (2, 1, 3) \qquad (1, 2, 3) \overset{\sigma_6}{\mapsto} (3, 2, 1)$$

In general, for the set $S = \{1, 2, \ldots, n\}$, there are $n!$ (n factorial) possible permutations. ∎

To *transpose* two adjacent elements simply means to interchange them; for example, the **transposition** (or **inversion**) of the pair 2, 3 is the pair 3, 2. *Every permutation can be obtained by a sequence of transpositions.* For example, consider the permutation σ_5 of $S = \{1, 2, 3\}$ defined in Example 1 above. The result of this permutation can be achieved by two successive transpositions of the original set:

$$(1, 2, 3) \xrightarrow[\text{transpose} \atop \text{2 and 3}]{} (1, 3, 2) \xrightarrow[\text{transpose} \atop \text{1 and 3}]{} (3, 1, 2)$$

Three transpositions are needed to give the permutation σ_6 of Example 1:

$$(1, 2, 3) \xrightarrow[\text{1 and 2}]{\text{transpose}} (2, 1, 3)$$

$$\xrightarrow[\text{1 and 3}]{\text{transpose}} (2, 3, 1)$$

$$\xrightarrow[\text{2 and 3}]{\text{transpose}} (3, 2, 1)$$

The number of transpositions needed to recover a given permutation is not unique. For example, you could always intersperse two successive transpositions, the second one of which simply undoes the first. However, what *is* unique is whether the number of transpositions is *even* or *odd*. If the number of transpositions that define a permutation is even, then the permutation is said to be **even**, and its **sign** is **+1**. If the number of transpositions that define a permutation is odd, then the permutation is said to be **odd**, and its **sign** is **−1**. The notation is as follows:

$$\operatorname{sgn} \sigma = \begin{cases} +1 & \text{if } \sigma \text{ is even} \\ -1 & \text{if } \sigma \text{ is odd} \end{cases}$$

Note that sgn σ can be defined as $(-1)^t$, where t is the number of transpositions that give σ.

Example 2: Determine the sign of the following permutation of the set $S = \{1, 2, 3, 4\}$:

$$\sigma: \quad (1, 2, 3, 4) \mapsto (3, 4, 1, 2)$$

The "brute-force" method is to explicitly determine the number of transpositions:

$$(1,\ 2,\ 3,\ 4) \rightarrow (1,\ 3,\ 2,\ 4) \rightarrow (3,\ 1,\ 2,\ 4)$$

$$\rightarrow (3,\ 1,\ 4,\ 2) \rightarrow (3,\ 4,\ 1,\ 2)$$

Since σ can be achieved by 4 successive transpositions, σ is even, so its sign is +1.

A faster method proceeds as follows: Determine how many pairs within the permutation have the property that a larger number precedes a smaller one. For example, in the permutation (3, 4, 1, 2) there are four such pairs: 3 precedes 1, 3 precedes 2, 4 precedes 1, and 4 precedes 2. The fact that the number of such pairs is even means the permutation itself is even, and its sign is +1. [Note: The number of pairs of elements that have the property that a larger number precedes a smaller one is the minimum number of transpositions that define the permutation. For example, since this number is four for the permutation (3, 4, 1, 2), at least four transpositions are needed to convert (1, 2, 3, 4) into (3, 4, 1, 2); the specific sequence of these four transpositions is shown above.] ∎

For every integer $n \geq 2$, the total number of permutations, $n!$, of the set $S = \{1, 2, \ldots, n\}$ is even. Exactly half of these permutations are even; the other half are odd.

Example 3: For the 6 = 3! permutations of the set $S = \{1, 2, 3\}$ given in Example 1, verify that the three permutations

$$\sigma_1, \sigma_4, \text{ and } \sigma_5 \text{ are even}$$

and, therefore, each has sign +1, while the other three permutations,

$$\sigma_2, \sigma_3, \text{ and } \sigma_6 \text{ are odd}$$

and each has sign −1. ∎

Now that the concepts of a permutation and its sign have been defined, the definition of the determinant of a matrix can be given. Let $A = [a_{ij}]$ be an n by n matrix, and let S_n denote the collection of *all* permutations of the set $S = \{1, 2, \ldots, n\}$. The **determinant** of A is defined to be the following sum:

$$\det A = \sum_{\sigma \in S_n} (\operatorname{sgn}\sigma) \cdot a_{1\sigma(1)} a_{2\sigma(2)} \cdots a_{n\sigma(n)} \qquad (*)$$

Example 4: Use definition (*) to derive an expression for the determinant of the general 2 by 2 matrix

$$A = \begin{bmatrix} a_{11} & a_{12} \\ a_{21} & a_{22} \end{bmatrix}$$

Since $n = 2$, there are 2! = 2 permutations of the set $\{1, 2\}$, namely,

$$(1, 2) \overset{\sigma_1}{\mapsto} (1, 2) \quad \text{and} \quad (1, 2) \overset{\sigma_2}{\mapsto} (2, 1)$$

The identity permutation, σ_1, is (always) even, so sgn σ_1 = +1, and the permutation σ_2 is odd, so sgn σ_2 = −1. Therefore, the sum (*) becomes

$$\det A = (\text{sgn}\,\sigma_1) \cdot a_{1\sigma_1(1)} a_{2\sigma_1(2)} + (\text{sgn}\,\sigma_2) \cdot a_{1\sigma_2(1)} a_{2\sigma_2(2)}$$
$$= (+1)a_{11}a_{22} + (-1)a_{12}a_{21}$$
$$= a_{11}a_{22} - a_{12}a_{21}$$

This formula is one you should memorize: To obtain the determinant of a 2 by 2 matrix, subtract the product of the off-diagonal entries from the product of the diagonal entries:

$$\det \begin{bmatrix} a_{11} & a_{12} \\ a_{21} & a_{22} \end{bmatrix} = \begin{vmatrix} a_{11} & a_{12} \\ a_{21} & a_{22} \end{vmatrix} = a_{11}a_{22} - a_{12}a_{21}$$

To illustrate,

$$\det \begin{bmatrix} 1 & 2 \\ 3 & 4 \end{bmatrix} = (1)(4) - (2)(3) = -2 \qquad \blacksquare$$

Example 5: Use definition (*) to derive an expression for the determinant of the general 3 by 3 matrix

$$A = \begin{bmatrix} a_{11} & a_{12} & a_{13} \\ a_{21} & a_{22} & a_{23} \\ a_{31} & a_{32} & a_{33} \end{bmatrix}$$

Since $n = 3$, there are $3! = 6$ permutations of $\{1, 2, 3\}$, and, therefore, six terms in the sum (*):

$$\det A = (\text{sgn}\,\sigma_1) \cdot a_{1\sigma_1(1)} a_{2\sigma_1(2)} a_{3\sigma_1(3)}$$
$$+ (\text{sgn}\,\sigma_2) \cdot a_{1\sigma_2(1)} a_{2\sigma_2(2)} a_{3\sigma_2(3)}$$
$$+ (\text{sgn}\,\sigma_3) \cdot a_{1\sigma_3(1)} a_{2\sigma_3(2)} a_{3\sigma_3(3)}$$
$$+ (\text{sgn}\,\sigma_4) \cdot a_{1\sigma_4(1)} a_{2\sigma_4(2)} a_{3\sigma_4(3)}$$
$$+ (\text{sgn}\,\sigma_5) \cdot a_{1\sigma_5(1)} a_{2\sigma_5(2)} a_{3\sigma_5(3)}$$
$$+ (\text{sgn}\,\sigma_6) \cdot a_{1\sigma_6(1)} a_{2\sigma_6(2)} a_{3\sigma_6(3)}$$

Using the notation for these permutations given in Example 1, as well as the evaluation of their signs in Example 3, the sum above becomes

$$\det A = (+1)a_{11}a_{22}a_{33} + (-1)a_{11}a_{23}a_{32}$$
$$+ (-1)a_{12}a_{21}a_{33} + (+1)a_{12}a_{23}a_{31}$$
$$+ (+1)a_{13}a_{21}a_{32} + (-1)a_{13}a_{22}a_{31}$$

or, more simply,

$$\det A = a_{11}a_{22}a_{33} + a_{12}a_{23}a_{31} + a_{13}a_{21}a_{32} \qquad (**)$$
$$- a_{11}a_{23}a_{32} - a_{12}a_{21}a_{33} - a_{13}a_{22}a_{31}$$

As you can see, there is quite a bit of work involved in computing a determinant of an n by n matrix directly from definition (*), particularly for large n. In applying the definition to evaluate the determinant of a 7 by 7 matrix, for example, the sum (*) would contain more than five *thousand* terms. This is why no one ever actually evaluates a determinant by this laborious method. ■

A simple way to produce the expansion (**) for the determinant of a 3 by 3 matrix is first to copy the first and second columns and place them after the matrix as follows:

$$\begin{bmatrix} a_{11} & a_{12} & a_{13} \\ a_{21} & a_{22} & a_{23} \\ a_{31} & a_{32} & a_{33} \end{bmatrix} \begin{matrix} a_{11} & a_{12} \\ a_{21} & a_{22} \\ a_{31} & a_{32} \end{matrix}$$

Then, multiply down along the three diagonals that start with the first row of the original matrix, and multiply up along the three diagonals that start with the bottom row of the original matrix. Keep the signs of the three "down" products, reverse the signs of the three "up" products, and add all six resulting terms; this gives (**). Note: This method works *only* for 3 by 3 matrices.

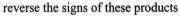

reverse the signs of these products

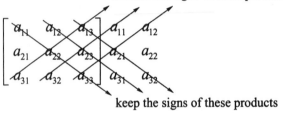

keep the signs of these products

Here's a helpful way to interpret definition (*). Note that in each of the products involved in the sum

$$\det A = \sum_{\sigma \in S_n} (\operatorname{sgn} \sigma) \cdot a_{1\sigma(1)} a_{2\sigma(2)} \cdots a_{n\sigma(n)}$$

there are n factors, no two of which come from the same row or column, a consequence of the bijectivity of every permutation. Using the 3 by 3 case above as a specific example, each of the six terms in the sum (**) can be illustrated as follows:

$$a_{11}a_{22}a_{33} \leftrightarrow \begin{bmatrix} \boxed{a_{11}} & a_{12} & a_{13} \\ a_{21} & \boxed{a_{22}} & a_{23} \\ a_{31} & a_{32} & \boxed{a_{33}} \end{bmatrix}$$

$$a_{12}a_{23}a_{31} \leftrightarrow \begin{bmatrix} a_{11} & \boxed{a_{12}} & a_{13} \\ a_{21} & a_{22} & \boxed{a_{23}} \\ \boxed{a_{31}} & a_{32} & a_{33} \end{bmatrix}$$

$$a_{13}a_{21}a_{32} \leftrightarrow \begin{bmatrix} a_{11} & a_{12} & \boxed{a_{13}} \\ \boxed{a_{21}} & a_{22} & a_{23} \\ a_{31} & \boxed{a_{32}} & a_{33} \end{bmatrix}$$

$$(-)a_{13}a_{22}a_{31} \leftrightarrow \begin{bmatrix} a_{11} & a_{12} & \boxed{a_{13}} \\ a_{21} & \boxed{a_{22}} & a_{23} \\ \boxed{a_{31}} & a_{32} & a_{33} \end{bmatrix}$$

$$(-)a_{12}a_{21}a_{33} \leftrightarrow \begin{bmatrix} a_{11} & \boxed{a_{12}} & a_{13} \\ \boxed{a_{21}} & a_{22} & a_{23} \\ a_{31} & a_{32} & \boxed{a_{33}} \end{bmatrix}$$

$$(-)a_{11}a_{23}a_{32} \leftrightarrow \begin{bmatrix} \boxed{a_{11}} & a_{12} & a_{13} \\ a_{21} & a_{22} & \boxed{a_{23}} \\ a_{31} & \boxed{a_{32}} & a_{33} \end{bmatrix}$$

These six products account for all possible ways of choosing three entries, no two of which reside in the same row or column. In general, then, the determinant is the sum of all possible products of n factors, no two of which come from the same row or column of the matrix, with the sign of each product, $a_{1j_1} a_{2j_2} \cdots a_{nj_n}$, determined by the sign of the corresponding permutation $\sigma : (1, 2, \ldots, n) \mapsto (j_1, j_2, \ldots, j_n)$.

Method 2 for defining the determinant. The second definition for the determinant follows from stating certain properties that the determinant function is to satisfy, which, it turns out, uniquely define the function. These properties will then lead to an *efficient* method for actually computing the determinant of a given matrix.

There exists a unique real-valued function—the **determinant function** (denoted **det**)—which is defined for n by n matrices and satisfies the following three properties:

Property 1: The determinant of a matrix is linear in each row.

Property 2: The determinant reverses sign if two rows are interchanged.

Property 3: The determinant of the identity matrix is equal to 1.

Property 1 deserves some explanation. Linearity of a function f means that $f(x+y) = f(x) + f(y)$ and, for any scalar k, $f(kx) = kf(x)$. Linearity of the determinant function in each row means, for example, that

$$\det \begin{bmatrix} - & \mathbf{r}_1 + \mathbf{r}_1' & - \\ - & \mathbf{r}_2 & - \\ & \vdots & \\ - & \mathbf{r}_n & - \end{bmatrix} = \det \begin{bmatrix} - & \mathbf{r}_1 & - \\ - & \mathbf{r}_2 & - \\ & \vdots & \\ - & \mathbf{r}_n & - \end{bmatrix} + \det \begin{bmatrix} - & \mathbf{r}_1' & - \\ - & \mathbf{r}_2 & - \\ & \vdots & \\ - & \mathbf{r}_n & - \end{bmatrix}$$

and

$$\det \begin{bmatrix} - & k\mathbf{r}_1 & - \\ - & \mathbf{r}_2 & - \\ & \vdots & \\ - & \mathbf{r}_n & - \end{bmatrix} = k \cdot \det \begin{bmatrix} - & \mathbf{r}_1 & - \\ - & \mathbf{r}_2 & - \\ & \vdots & \\ - & \mathbf{r}_n & - \end{bmatrix}$$

Although these two equations illustrate linearity in the *first* row, linearity of the determinant function can be applied to *any* row.

Property 2 can be used to derive another important property of the determinant function:

Property 4: The determinant of a matrix with two identical rows is equal to 0.

The proof of this fact is easy: Assume that for the matrix A, Row i = Row j. By interchanging these two rows, the determinant changes sign (by Property 2). However, since these two rows are the same, interchanging them obviously leaves the matrix and, therefore, the determinant unchanged. Since 0 is the only number which equals its own opposite, det $A = 0$.

One of the most important matrix operations is adding a multiple of one row to another row. How the determinant reacts to this operation is a key property in evaluating it:

Property 5: Adding a multiple of one row to another row leaves the determinant unchanged.

The idea of the general proof will be illustrated by the following specific illustration. Suppose the matrix A is 4 by 4, and k times Row 2 is added to Row 3:

$$A = \begin{bmatrix} - & \mathbf{r}_1 & - \\ - & \mathbf{r}_2 & - \\ - & \mathbf{r}_3 & - \\ - & \mathbf{r}_4 & - \end{bmatrix} \longrightarrow \begin{bmatrix} - & \mathbf{r}_1 & - \\ - & \mathbf{r}_2 & - \\ - & \mathbf{r}_3 + k\mathbf{r}_2 & - \\ - & \mathbf{r}_4 & - \end{bmatrix} = A'$$

By linearity applied to the third row,

$$\det A' = \det \begin{bmatrix} - & \mathbf{r}_1 & - \\ - & \mathbf{r}_2 & - \\ - & \mathbf{r}_3 + k\mathbf{r}_2 & - \\ - & \mathbf{r}_4 & - \end{bmatrix}$$

$$= \det \begin{bmatrix} - & \mathbf{r}_1 & - \\ - & \mathbf{r}_2 & - \\ - & \mathbf{r}_3 & - \\ - & \mathbf{r}_4 & - \end{bmatrix} + \det \begin{bmatrix} - & \mathbf{r}_1 & - \\ - & \mathbf{r}_2 & - \\ - & k\mathbf{r}_2 & - \\ - & \mathbf{r}_4 & - \end{bmatrix}$$

$$= \det \begin{bmatrix} - & \mathbf{r}_1 & - \\ - & \mathbf{r}_2 & - \\ - & \mathbf{r}_3 & - \\ - & \mathbf{r}_4 & - \end{bmatrix} + k \det \begin{bmatrix} - & \mathbf{r}_1 & - \\ - & \mathbf{r}_2 & - \\ - & \mathbf{r}_2 & - \\ - & \mathbf{r}_4 & - \end{bmatrix}$$

But the second term in this last equation is zero, because the matrix contains two identical rows (Property 4). Therefore,

$$\det A' = \det \begin{bmatrix} - & \mathbf{r}_1 & - \\ - & \mathbf{r}_2 & - \\ - & \mathbf{r}_3 + k\mathbf{r}_2 & - \\ - & \mathbf{r}_4 & - \end{bmatrix} = \det \begin{bmatrix} - & \mathbf{r}_1 & - \\ - & \mathbf{r}_2 & - \\ - & \mathbf{r}_3 & - \\ - & \mathbf{r}_4 & - \end{bmatrix} = \det A$$

The purpose of adding a multiple of one row to another row is to simplify a matrix (when solving a linear system, for example). For a square matrix, the goal of these operations is to reduce the given matrix to an upper triangular one. So the natural question at this point is: What is the determinant of an upper triangular matrix?

Property 6: The determinant of an upper triangular (or diagonal) matrix is equal to the product of the diagonal entries.

To prove this property, assume that the given matrix A has been reduced to upper triangular form by adding multiples of rows to other rows *and* assume that none of the resulting diagonal entries is equal to 0. (The case of a 0 diagonal entry will be discussed later.) This upper triangular matrix can be transformed into a *diagonal* one by adding multiples of lower rows to higher ones. (Recall Example 10, page 97.) At each step of this transformation, the determinant is left unchanged, by Property 5. Therefore, the problem of evaluating the determinant of the original matrix has been reduced to evaluating the determinant of an upper triangular matrix, which in turn has been reduced to evaluating the determinant of a diagonal matrix. By factoring out each diagonal entry and using Property 1 (linearity in each row), Property 3 (det I = 1) gives the desired result:

$$\det\begin{bmatrix} a_{11} & & & \\ & a_{22} & & \\ & & \ddots & \\ & & & a_{nn} \end{bmatrix} = a_{11}\det\begin{bmatrix} 1 & & & \\ & a_{22} & & \\ & & \ddots & \\ & & & a_{nn} \end{bmatrix}$$

$$= a_{11}a_{22}\det\begin{bmatrix} 1 & & & \\ & 1 & & \\ & & \ddots & \\ & & & a_{nn} \end{bmatrix}$$

$$\vdots$$

$$= a_{11}a_{22}\cdots a_{nn}\det\begin{bmatrix} 1 & & & \\ & 1 & & \\ & & \ddots & \\ & & & 1 \end{bmatrix}$$

$$= a_{11}a_{22}\cdots a_{nn}$$

Now, to handle the case of a zero diagonal entry, the following property will be established:

Property 7:   A matrix with a row of zeros has determinant zero.

This is also easy to prove. As in the proof of Property 5, the essential idea of this proof will also be illustrated by a specific example. Consider the 3 by 3 matrix

$$A = \begin{bmatrix} * & * & * \\ 0 & 0 & 0 \\ * & * & * \end{bmatrix}$$

(Recall that each * indicates an entry whose value is irrelevant to the present discussion.)

Since for any scalar k,

$$A = \begin{bmatrix} * & * & * \\ 0 & 0 & 0 \\ * & * & * \end{bmatrix} = \begin{bmatrix} * & * & * \\ k \cdot 0 & k \cdot 0 & k \cdot 0 \\ * & * & * \end{bmatrix}$$

linearity of the determinant implies

$$\det \begin{bmatrix} * & * & * \\ 0 & 0 & 0 \\ * & * & * \end{bmatrix} = \det \begin{bmatrix} * & * & * \\ k \cdot 0 & k \cdot 0 & k \cdot 0 \\ * & * & * \end{bmatrix} = k \det \begin{bmatrix} * & * & * \\ 0 & 0 & 0 \\ * & * & * \end{bmatrix}$$

But, if det A is equal to k det A for any scalar k, then det A must be 0.

Now, to complete the discussion of Property 6: If a diagonal entry in an upper triangular matrix is equal to 0, then the process of adding a multiple of one row to another can produce a row of zeros. For example,

$$\begin{bmatrix} 1 & -3 & 5 \\ 0 & 0 & 2 \\ 0 & 0 & 3 \end{bmatrix} \xrightarrow{-\frac{3}{2}\mathbf{r}_2 \text{ added to } \mathbf{r}_3} \begin{bmatrix} 1 & -3 & 5 \\ 0 & 0 & 2 \\ 0 & 0 & 0 \end{bmatrix}$$

This step does not change the determinant (Property 3), so the determinant of the original matrix is equal to the determinant of a matrix with a row of zeros, which is zero (Property 4). But in this case at least one of the diagonal entries of the upper triangular matrix is 0, so the determinant does indeed equal the product of the diagonal entries. Generalizing these arguments fully establishes Property 6.

Example 6: Evaluate the determinant of

$$A = \begin{bmatrix} 1 & -3 & 7 \\ 2 & 2 & 1 \\ -4 & -4 & 3 \end{bmatrix}$$

Reduce the matrix to an upper triangular one,

$$\begin{bmatrix} 1 & -3 & 7 \\ 2 & 2 & 1 \\ -4 & -4 & 3 \end{bmatrix} \xrightarrow[4r_1 \text{ added to } r_3]{-2r_1 \text{ added to } r_2} \begin{bmatrix} 1 & -3 & 7 \\ 0 & 8 & -13 \\ 0 & -16 & 31 \end{bmatrix}$$

$$\xrightarrow{2r_2 \text{ added to } r_3} \begin{bmatrix} 1 & -3 & 7 \\ 0 & 8 & -13 \\ 0 & 0 & 5 \end{bmatrix}$$

in order to exploit Property 6—that none of these operations changes the determinant—and Property 7—that the determinant of an upper triangular matrix is equal to the product of the diagonal entries. The result is

$$\det A = \det \begin{bmatrix} 1 & -3 & 7 \\ 0 & 8 & -13 \\ 0 & 0 & 5 \end{bmatrix} = (1)(8)(5) = 40$$

∎

Example 7: Evaluate the determinant of

$$A = \begin{bmatrix} 2 & 1 & 0 & -1 \\ 1 & 0 & -1 & 2 \\ 0 & -1 & 2 & 1 \\ -1 & 2 & 1 & 0 \end{bmatrix}$$

The following elementary row operations reduce A to an upper triangular matrix:

$$\begin{bmatrix} 2 & 1 & 0 & -1 \\ 1 & 0 & -1 & 2 \\ 0 & -1 & 2 & 1 \\ -1 & 2 & 1 & 0 \end{bmatrix} \xrightarrow{\ r_1 \leftrightarrow r_2\ } \begin{bmatrix} 1 & 0 & -1 & 2 \\ 2 & 1 & 0 & -1 \\ 0 & -1 & 2 & 1 \\ -1 & 2 & 1 & 0 \end{bmatrix}$$

$$\xrightarrow[\ r_1 \text{ added to } r_4\ ]{-2r_1 \text{ added to } r_2} \begin{bmatrix} 1 & 0 & -1 & 2 \\ 0 & 1 & 2 & -5 \\ 0 & -1 & 2 & 1 \\ 0 & 2 & 0 & 2 \end{bmatrix}$$

$$\xrightarrow[\ -2r_2 \text{ added to } r_4\ ]{r_2 \text{ added to } r_3} \begin{bmatrix} 1 & 0 & -1 & 2 \\ 0 & 1 & 2 & -5 \\ 0 & 0 & 4 & -4 \\ 0 & 0 & -4 & 12 \end{bmatrix}$$

$$\xrightarrow{\ r_3 \text{ added to } r_4\ } \begin{bmatrix} 1 & 0 & -1 & 2 \\ 0 & 1 & 2 & -5 \\ 0 & 0 & 4 & -4 \\ 0 & 0 & 0 & 8 \end{bmatrix}$$

None of these operations alters the determinant, except for the row exchange in the first step, which reverses its sign. Since the determinant of the final upper triangular matrix is $(1)(1)(4)(8) = 32$, the determinant of the original matrix A is -32. ∎

Example 8: Let C be a square matrix. What does the rank of C say about its determinant?

Let C be $n \times n$ and first assume that the rank of C is less than n. This means that if C is reduced to echelon form by a sequence of elementary row operations, at least one row of zeros appears at the bottom of the reduced matrix. But a square matrix with a row of zeros has determinant zero. Since no elementary row operation can turn a nonzero-determinant matrix into a zero-determinant one, the original matrix C had to have determinant zero also.

On the other hand, if rank $C = n$, then all the rows are independent, and the echelon form of C will be upper triangular with no zeros on the diagonal. Thus, the determinant of the reduced matrix is nonzero. Since no elementary row operation can transform a zero-determinant matrix into a nonzero-determinant one, the original matrix C had to have a nonzero determinant. To summarize then,

$$\text{If } C \text{ is } n \times n, \begin{cases} \text{rank } C < n & \Leftrightarrow & \det C = 0 \\ \text{rank } C = n & \Leftrightarrow & \det C \neq 0 \end{cases} \qquad \blacksquare$$

Example 9: Evaluate the determinant of

$$A = \begin{bmatrix} 1 & 2 & 3 \\ 4 & 5 & 6 \\ 7 & 8 & 9 \end{bmatrix}$$

None of the following row operations affects the determinant of A:

$$\begin{bmatrix} 1 & 2 & 3 \\ 4 & 5 & 6 \\ 7 & 8 & 9 \end{bmatrix} \xrightarrow[\begin{subarray}{c} -4\mathbf{r}_1 \text{ added to } \mathbf{r}_2 \\ -7\mathbf{r}_1 \text{ added to } \mathbf{r}_3 \end{subarray}]{} \begin{bmatrix} 1 & 2 & 3 \\ 0 & -3 & -6 \\ 0 & -6 & -12 \end{bmatrix}$$

$$\xrightarrow[-2\mathbf{r}_2 \text{ added to } \mathbf{r}_3]{} \begin{bmatrix} 1 & 2 & 3 \\ 0 & -3 & -6 \\ 0 & 0 & 0 \end{bmatrix}$$

Because this final matrix has a zero row, its determinant is zero, which implies $\det A = 0$. ∎

Example 10: What is the rank of the following matrix?

$$A = \begin{bmatrix} 1 & 2 & 3 \\ 4 & 5 & 6 \\ 7 & 8 & 9 \end{bmatrix}$$

Since the third row is a linear combination, $\mathbf{r}_3 = -\mathbf{r}_1 + 2\mathbf{r}_2$, of the first two rows, a row of zeros results when A is reduced to echelon form, as in Example 9 above. Since just 2 nonzero rows remain, rank $A = 2$. ∎

The three preceding examples illustrate the following important theorem:

Theorem E. Consider a collection $\{\mathbf{v}_1, \mathbf{v}_2, \ldots, \mathbf{v}_n\}$ of n vectors from $\mathbf{R}^n$. Then this collection is linearly independent if and only if the determinant of the matrix whose rows are $\mathbf{v}_1, \mathbf{v}_2, \ldots, \mathbf{v}_n$ is not zero.

In fact, Theorem E can be amended: If a collection of n vectors from $\mathbf{R}^n$ is linearly independent, then it also spans $\mathbf{R}^n$ (and conversely); therefore, the collection is a basis for $\mathbf{R}^n$.

Example 11: Let A be a real 5 by 5 matrix such that the sum of the entries in each row is zero. What can you say about the determinant of A?

Solution 1. The equation $x_1 + x_2 + x_3 + x_4 + x_5 = 0$ describes a 4-dimensional subspace of $\mathbf{R}^5$, since every point in this subspace has the form

$$(x_1, x_2, x_3, x_4, -x_1 - x_2 - x_3 - x_4)$$

which contains 4 independent parameters. Since every row of the matrix A has this form, A contains 5 vectors all lying in a 4-dimensional subspace. Since such a space can contain at most 4 linearly independent vectors, the 5 row vectors of A must be dependent. Thus, det $A = 0$.

Solution 2. If $\mathbf{x}_0$ is the column vector $(1, 1, 1, 1, 1)^T$, then the product $A\mathbf{x}_0$ equals the zero vector. Since the homogeneous system $A\mathbf{x} = \mathbf{0}$ has a nontrivial solution, A must have determinant zero (Theorem G, page 239). ∎

Example 12: Do the matrices in $M_{2\times2}(\mathbf{R})$ with determinant 1 form a subspace of $M_{2\times2}(\mathbf{R})$?

No. The determinant function is incompatible with the usual vector space operations: The set of 2×2 matrices with determinant 1 is not closed under addition or scalar multiplication, and, therefore, cannot form a subspace of $M_{2\times2}(\mathbf{R})$. A counterexample to closure under addition is provided by the matrices I and $-I$; although each has determinant 1, their sum, $I + (-I) = 0$, clearly does not. ∎

Example 13: Given that

$$\det\begin{bmatrix} 1 & -3 & 7 \\ 2 & 2 & 1 \\ -4 & -4 & 3 \end{bmatrix} = 40$$

(see Example 6), compute the determinant of the matrix

$$\begin{bmatrix} 2 & -6 & 14 \\ 4 & 4 & 2 \\ -8 & -8 & 6 \end{bmatrix}$$

obtained by multiplying every entry of the first matrix by 2.

This question is asking for $\det(2A)$ in terms of det A. If just one row of A were multiplied by 2, the determinant would be multiplied by 2, by Property 1 above. But, in this case, all three rows have been multiplied by 2, so the determinant is multiplied by three factors of 2:

$$\det(2A) = 2^3 \det A$$

This gives $\det(2A) = 8 \cdot 40 = 320$. In general, if A is an n by n matrix and k is a scalar, then

$$\det(kA) = k^n \det A \quad ∎$$

Example 14: If A and B are square matrices of the same size, is the equation det $(A + B)$ = det A + det B always true?

Let A and B be the following 2 by 2 matrices:

$$A = \begin{bmatrix} 1 & 2 \\ 3 & 4 \end{bmatrix} \quad \text{and} \quad B = \begin{bmatrix} 5 & 6 \\ 7 & 8 \end{bmatrix}$$

Then det A = det B = –2, but

$$\det(A + B) = \det \begin{bmatrix} 6 & 8 \\ 10 & 12 \end{bmatrix} = -8 \neq -4 = \det A + \det B$$

Thus, det $(A + B)$ = det A + det B is not an identity. [Note: This does not mean that this equation never holds. It certainly *is* an identity for 1×1 matrices, and, making just one change in the entries of the matrices above (namely, changing the entry b_{22} from 8 to 12),

$$A = \begin{bmatrix} 1 & 2 \\ 3 & 4 \end{bmatrix} \quad \text{and} \quad B = \begin{bmatrix} 5 & 6 \\ 7 & 12 \end{bmatrix}$$

yields a pair of matrices that *does* satisfy det $(A + B)$ = det A + det B, as you may check.] ■

Example 15: One of the most important properties of the determinant function is that the determinant of the product of two square matrices (of the same size) is equal to the product of the individual determinants. That is,

$$\boxed{\det(AB) = (\det A)(\det B)}$$

is an identity for all matrices A and B for which both sides are defined.

THE
DETERMINANT

(a) Verify this identity for the matrices

$$A = \begin{bmatrix} 2 & -1 \\ -3 & 5 \end{bmatrix} \quad \text{and} \quad B = \begin{bmatrix} -6 & -2 \\ 4 & 3 \end{bmatrix}$$

(b) Assuming that A is an invertible matrix, what is the relationship between the determinant of A and the determinant of A^{-1}?

(c) If A is a square matrix and k is an integer greater than 1, what relationship exists between det (A^k) and det A?

The solutions are as follows:

(a) It is easy to see that det $A = 7$ and det $B = -10$. The product of A and B,

$$AB = \begin{bmatrix} 2 & -1 \\ -3 & 5 \end{bmatrix}\begin{bmatrix} -6 & -2 \\ 4 & 3 \end{bmatrix} = \begin{bmatrix} -16 & -7 \\ 38 & 21 \end{bmatrix}$$

has determinant $(-16)(21) - (38)(-7) = -336 + 266 = -70$. Thus,

$$\text{det } (AB) = -70 = (7)(-10) = (\text{det } A)(\text{det } B)$$

as expected.

(b) Taking the determinant of both sides of the equation $AA^{-1} = I$ yields

$$\det(AA^{-1}) = \det(I)$$
$$(\det A)(\det A^{-1}) = 1$$
$$\det A^{-1} = (\det A)^{-1}$$

Note that the identity $(\det A)(\det A^{-1}) = 1$ implies that a necessary condition for A^{-1} to exist is that det A is nonzero. (In fact, this condition is also sufficient; see Theorem H, page 243.)

(c) Let $k = 2$; then det $(A^2) = $ det $(AA) = $ (det A)(det A) = (det $A)^2$. If $k = 3$, then det $(A^3) = $ det $(A^2A) = $ det (A^2)(det A) = (det $A)^2$(det A) = (det $A)^3$. The pattern is clear: det $(A^k) = $ (det $A)^k$. [You may find it instructive to give a more rigorous proof of this statement by a straightforward induction argument.] ■

Laplace Expansions for the Determinant

Using the definition of the determinant, the following expression was derived in Example 5:

$$\begin{vmatrix} a_{11} & a_{12} & a_{13} \\ a_{21} & a_{22} & a_{23} \\ a_{31} & a_{32} & a_{33} \end{vmatrix} = a_{11}a_{22}a_{33} + a_{12}a_{23}a_{31} + a_{13}a_{21}a_{32} \\ - a_{11}a_{23}a_{32} - a_{12}a_{21}a_{33} - a_{13}a_{22}a_{31}$$

This equation can be rewritten as follows:

$$\begin{vmatrix} a_{11} & a_{12} & a_{13} \\ a_{21} & a_{22} & a_{23} \\ a_{31} & a_{32} & a_{33} \end{vmatrix} = a_{11}(a_{22}a_{33} - a_{23}a_{32}) + a_{12}(a_{23}a_{31} - a_{21}a_{33}) \\ + a_{13}(a_{21}a_{32} - a_{22}a_{31})$$

$$= a_{11}\begin{vmatrix} a_{22} & a_{23} \\ a_{32} & a_{33} \end{vmatrix} + a_{12}\left(-\begin{vmatrix} a_{21} & a_{23} \\ a_{31} & a_{33} \end{vmatrix} \right) + a_{13}\begin{vmatrix} a_{21} & a_{22} \\ a_{31} & a_{32} \end{vmatrix}$$

Each term on the right has the following form:

$$(\text{entry in the first row}) \cdot \pm \left(\begin{array}{l} \text{determinant of the matrix that} \\ \text{remains when the row and} \\ \text{column containing that entry are} \\ \text{removed from the original matrix} \end{array} \right)$$

In particular, note that

$$\begin{bmatrix} \boxed{a_{11}} & a_{12} & a_{13} \\ a_{21} & a_{22} & a_{23} \\ a_{31} & a_{32} & a_{33} \end{bmatrix} \text{ gives the term } a_{11} \begin{vmatrix} a_{22} & a_{23} \\ a_{32} & a_{33} \end{vmatrix}$$

$$\begin{bmatrix} a_{11} & \boxed{a_{12}} & a_{13} \\ a_{21} & a_{22} & a_{23} \\ a_{31} & a_{32} & a_{33} \end{bmatrix} \text{ gives the term } a_{12}\left(-\begin{vmatrix} a_{21} & a_{23} \\ a_{31} & a_{33} \end{vmatrix} \right)$$

$$\begin{bmatrix} a_{11} & a_{12} & \boxed{a_{13}} \\ a_{21} & a_{22} & a_{23} \\ a_{31} & a_{32} & a_{33} \end{bmatrix} \text{ gives the term } a_{13} \begin{vmatrix} a_{21} & a_{22} \\ a_{31} & a_{32} \end{vmatrix}$$

If $A = [a_{ij}]$ is an $n \times n$ matrix, then the determinant of the $(n - 1) \times (n - 1)$ matrix that remains once the row and column containing the entry a_{ij} are deleted is called the a_{ij} **minor**, denoted $\text{mnr}(a_{ij})$. If the a_{ij} minor is multiplied by $(-1)^{i+j}$, the result is called the a_{ij} **cofactor**, denoted $\text{cof}(a_{ij})$. That is,

$$\text{cof}(a_{ij}) = (-1)^{i+j} \cdot \text{mnr}(a_{ij})$$

Using this terminology, the equation given above for the determinant of the 3×3 matrix A is equal to the sum of the products of the entries in the first row and their cofactors:

$$\begin{vmatrix} a_{11} & a_{12} & a_{13} \\ a_{21} & a_{22} & a_{23} \\ a_{31} & a_{32} & a_{33} \end{vmatrix} = a_{11}\,\text{cof}(a_{11}) + a_{12}\,\text{cof}(a_{12}) + a_{13}\,\text{cof}(a_{13})$$

This is called the **Laplace expansion** by the first row. It can also be shown that the determinant is equal to the Laplace expansion by the *second* row,

$$\begin{vmatrix} a_{11} & a_{12} & a_{13} \\ a_{21} & a_{22} & a_{23} \\ a_{31} & a_{32} & a_{33} \end{vmatrix} = a_{21}\cof(a_{21}) + a_{22}\cof(a_{22}) + a_{23}\cof(a_{23})$$

or by the *third* row,

$$\begin{vmatrix} a_{11} & a_{12} & a_{13} \\ a_{21} & a_{22} & a_{23} \\ a_{31} & a_{32} & a_{33} \end{vmatrix} = a_{31}\cof(a_{31}) + a_{32}\cof(a_{32}) + a_{33}\cof(a_{33})$$

Even more is true. The determinant is also equal to the Laplace expansion by the first *column*,

$$\begin{vmatrix} a_{11} & a_{12} & a_{13} \\ a_{21} & a_{22} & a_{23} \\ a_{31} & a_{32} & a_{33} \end{vmatrix} = a_{11}\cof(a_{11}) + a_{21}\cof(a_{21}) + a_{31}\cof(a_{31})$$

by the second column, or by the third column. Although the Laplace expansion formula for the determinant has been explicitly verified only for a 3×3 matrix and only for the first row, it can be proved that *the determinant of any $n \times n$ matrix is equal to the Laplace expansion by any row or any column.*

Example 16: Evaluate the determinant of the following matrix using the Laplace expansion by the second column:

$$A = \begin{bmatrix} 2 & -1 & -1 \\ -3 & 2 & 1 \\ 5 & 0 & -2 \end{bmatrix}$$

The entries in the second column are $a_{12} = -1$, $a_{22} = 2$, and $a_{32} = 0$. The minors of these entries, mnr(a_{12}), mnr(a_{22}), and mnr(a_{32}), are computed as follows:

$$\Rightarrow a_{12} \text{ minor} = \begin{vmatrix} -3 & 1 \\ 5 & -2 \end{vmatrix}$$
$$= (-3)(-2) - (5)(1) = 1$$

$$\Rightarrow a_{22} \text{ minor} = \begin{vmatrix} 2 & -1 \\ 5 & -2 \end{vmatrix}$$
$$= (2)(-2) - (5)(-1) = 1$$

$$\Rightarrow a_{32} \text{ minor} = \begin{vmatrix} 2 & -1 \\ -3 & 1 \end{vmatrix}$$
$$= (2)(1) - (-3)(-1) = -1$$

Since the cofactors of the second-column entries are

$$\text{cof}(a_{12}) = (-1)^{1+2} \text{mnr}(a_{12}) = -\text{mnr}(a_{12}) = -1$$
$$\text{cof}(a_{22}) = (-1)^{2+2} \text{mnr}(a_{22}) = \text{mnr}(a_{22}) = 1$$
$$\text{cof}(a_{32}) = (-1)^{3+2} \text{mnr}(a_{32}) = -\text{mnr}(a_{32}) = -(-1) = 1$$

the Laplace expansion by the second column becomes

$$\begin{aligned} \det A &= a_{12}\,\text{cof}(a_{12}) + a_{22}\,\text{cof}(a_{22}) + a_{32}\,\text{cof}(a_{32}) \\ &= (-1)(-1) + (2)(1) + (0)(1) \\ &= 3 \end{aligned}$$

Note that it was unnecessary to compute the minor or the co-factor of the (3, 2) entry in A, since that entry was 0. In general, then, when computing a determinant by the Laplace expansion method, choose the row or column with the most zeros. The minors of those entries need not be evaluated, because they will contribute nothing to the determinant. ∎

The factor $(-1)^{i+j}$ which multiplies the a_{ij} minor to give the a_{ij} cofactor leads to a checkerboard pattern of signs; each sign gives the value of this factor when computing the a_{ij} cofactor from the a_{ij} minor. For example, the checkerboard pattern for a 3×3 matrix looks like this:

$$\begin{bmatrix} + & - & + \\ - & + & - \\ + & - & + \end{bmatrix}$$

For a 4×4 matrix, the checkerboard has the form

$$\begin{bmatrix} + & - & + & - \\ - & + & - & + \\ + & - & + & - \\ - & + & - & + \end{bmatrix}$$

and so on.

Example 17: Compute the determinant of the following matrix:

$$A = \begin{bmatrix} 0 & 7 & 1 & -5 \\ -3 & 2 & -1 & 1 \\ 1 & 0 & 0 & 2 \\ 2 & -4 & -2 & 0 \end{bmatrix}$$

First, find the row or column with the most zeros. Here, it's the third row, which contains two zeros; the Laplace expansion by this row will contain only two nonzero terms. The checkerboard pattern displayed above for a 4 by 4 matrix implies that the minor of the entry $a_{31} = 1$ will be multiplied by $+1$, and the minor of the entry $a_{34} = 2$ will be multiplied by -1 to give the respective cofactors:

$$\begin{bmatrix} 0 & 7 & 1 & -5 \\ -3 & 2 & -1 & 1 \\ 1 & 0 & 0 & 2 \\ 2 & -4 & -2 & 0 \end{bmatrix} \Rightarrow \text{cof}(a_{31}) = + \begin{vmatrix} 7 & 1 & -5 \\ 2 & -1 & 1 \\ -4 & -2 & 0 \end{vmatrix}$$

$$\begin{bmatrix} 0 & 7 & 1 & -5 \\ -3 & 2 & -1 & 1 \\ 1 & 0 & 0 & 2 \\ 2 & -4 & -2 & 0 \end{bmatrix} \Rightarrow \text{cof}(a_{34}) = - \begin{vmatrix} 0 & 7 & 1 \\ -3 & 2 & -1 \\ 2 & -4 & -2 \end{vmatrix}$$

Now, each of these cofactors—which are themselves determinants—can be evaluated by a Laplace expansion. Expanding by the third column,

$$\text{cof}(a_{31}) = \begin{vmatrix} 7 & 1 & -5 \\ 2 & -1 & 1 \\ -4 & -2 & 0 \end{vmatrix} = -5\begin{vmatrix} 2 & -1 \\ -4 & -2 \end{vmatrix} + 1\left(-\begin{vmatrix} 7 & 1 \\ -4 & -2 \end{vmatrix}\right)$$

$$= -5(-4-4) - (-14+4)$$

$$= 50$$

The other cofactor is evaluated by expanding along its first row:

$$\text{cof}(a_{34}) = -\begin{vmatrix} 0 & 7 & 1 \\ -3 & 2 & -1 \\ 2 & -4 & -2 \end{vmatrix} = -\left[7\left(-\begin{vmatrix} -3 & -1 \\ 2 & -2 \end{vmatrix}\right) + 1\begin{vmatrix} -3 & 2 \\ 2 & -4 \end{vmatrix} \right]$$

$$= -\left[-7(6+2) + (12-4) \right]$$

$$= 48$$

Therefore, evaluating det A by the Laplace expansion along A's third row yields

$$\det A = a_{31}\,\text{cof}(a_{31}) + a_{34}\,\text{cof}(a_{34})$$

$$= (1)(50) + (2)(48)$$

$$= 146 \quad \blacksquare$$

Example 18: The cross product of two 3-vectors, $\mathbf{x} = x_1\mathbf{i} + x_2\mathbf{j} + x_3\mathbf{k}$ and $\mathbf{y} = y_1\mathbf{i} + y_2\mathbf{j} + y_3\mathbf{k}$, is most easily evaluated by performing the Laplace expansion along the first row of the symbolic determinant

$$\begin{vmatrix} \mathbf{i} & \mathbf{j} & \mathbf{k} \\ x_1 & x_2 & x_3 \\ y_1 & y_2 & y_3 \end{vmatrix}$$

This expansion gives

$$\mathbf{x} \times \mathbf{y} = \begin{vmatrix} \mathbf{i} & \mathbf{j} & \mathbf{k} \\ x_1 & x_2 & x_3 \\ y_1 & y_2 & y_3 \end{vmatrix} = \mathbf{i} \begin{vmatrix} x_2 & x_3 \\ y_2 & y_3 \end{vmatrix} - \mathbf{j} \begin{vmatrix} x_1 & x_3 \\ y_1 & y_3 \end{vmatrix} + \mathbf{k} \begin{vmatrix} x_1 & x_2 \\ y_1 & y_2 \end{vmatrix}$$

To illustrate, the cross product of the vectors $\mathbf{x} = 3\mathbf{j} - 3\mathbf{k}$ and $\mathbf{y} = -2\mathbf{i} + 2\mathbf{j} - \mathbf{k}$ is

$$\mathbf{x} \times \mathbf{y} = \begin{vmatrix} \mathbf{i} & \mathbf{j} & \mathbf{k} \\ 0 & 3 & -3 \\ -2 & 2 & -1 \end{vmatrix} = \mathbf{i} \begin{vmatrix} 3 & -3 \\ 2 & -1 \end{vmatrix} - \mathbf{j} \begin{vmatrix} 0 & -3 \\ -2 & -1 \end{vmatrix} + \mathbf{k} \begin{vmatrix} 0 & 3 \\ -2 & 2 \end{vmatrix}$$

$$= \mathbf{i}(-3+6) - \mathbf{j}(0-6) + \mathbf{k}(0+6)$$
$$= 3\mathbf{i} + 6\mathbf{j} + 6\mathbf{k}$$

This calculation was performed in Example 27, page 45. ■

Example 19: Is there a connection between the determinant of A^T and the determinant of A?

In the 2 by 2 case, it is easy to see that $\det(A^T) = \det A$:

$$\det A = \begin{vmatrix} a & b \\ c & d \end{vmatrix} = ad - bc$$

$$\det A^T = \begin{vmatrix} a & c \\ b & d \end{vmatrix} = ad - bc$$

In the 3 by 3 case, the Laplace expansion along the first row of A gives the same result as the Laplace expansion along the first column of A^T, implying that $\det(A^T) = \det A$:

$$\det A = \begin{vmatrix} a & b & c \\ d & e & f \\ g & h & i \end{vmatrix} = a \begin{vmatrix} e & f \\ h & i \end{vmatrix} - b \begin{vmatrix} d & f \\ g & i \end{vmatrix} + c \begin{vmatrix} d & e \\ g & h \end{vmatrix}$$

$$\det A^{\mathrm{T}} = \begin{vmatrix} a & d & g \\ b & e & h \\ c & f & i \end{vmatrix} = a \begin{vmatrix} e & h \\ f & i \end{vmatrix} - b \begin{vmatrix} d & g \\ f & i \end{vmatrix} + c \begin{vmatrix} d & g \\ e & h \end{vmatrix}$$

Starting with the expansion

$$\det A = \sum_{\sigma \in S_n} (\mathrm{sgn}\,\sigma) \cdot a_{1\sigma(1)} a_{2\sigma(2)} \cdots a_{n\sigma(n)}$$

for the determinant, it is not difficult to give a general proof that $\det (A^{\mathrm{T}}) = \det A$. ∎

Example 20: Apply the result $\det (A^{\mathrm{T}}) = \det A$ to evaluate

$$\begin{vmatrix} e & r & a \\ g & a & p \\ o & n & e \end{vmatrix}$$

given that

$$\begin{vmatrix} a & p & e \\ e & g & o \\ r & a & n \end{vmatrix} = 20$$

(where a, e, g, n, o, p, and r are scalars).

Since one row exchange reverses the sign of the determinant (Property 2), two row exchanges,

$$\begin{bmatrix} a & p & e \\ e & g & o \\ r & a & n \end{bmatrix} \xrightarrow{\ r_1 \leftrightarrow r_2\ } \begin{bmatrix} e & g & o \\ a & p & e \\ r & a & n \end{bmatrix} \xrightarrow{\ r_2 \leftrightarrow r_3\ } \begin{bmatrix} e & g & o \\ r & a & n \\ a & p & e \end{bmatrix}$$

will leave the determinant unchanged:

$$\det \begin{bmatrix} a & p & e \\ e & g & o \\ r & a & n \end{bmatrix} = \det \begin{bmatrix} e & g & o \\ r & a & n \\ a & p & e \end{bmatrix}$$

But the determinant of a matrix is equal to the determinant of its transpose, so

$$\det \begin{bmatrix} e & g & o \\ r & a & n \\ a & p & e \end{bmatrix} = \det \begin{bmatrix} e & g & o \\ r & a & n \\ a & p & e \end{bmatrix}^{\mathrm{T}} = \det \begin{bmatrix} e & r & a \\ g & a & p \\ o & n & e \end{bmatrix}$$

Therefore,

$$\begin{vmatrix} e & r & a \\ g & a & p \\ o & n & e \end{vmatrix} = \begin{vmatrix} a & p & e \\ e & g & o \\ r & a & n \end{vmatrix} = 20 \qquad \blacksquare$$

Example 21: Given that the numbers 1547, 2329, 3893, and 4471 are all divisible by 17, prove that the determinant of

$$A = \begin{bmatrix} 1 & 5 & 4 & 7 \\ 2 & 3 & 2 & 9 \\ 3 & 8 & 9 & 3 \\ 4 & 4 & 7 & 1 \end{bmatrix}$$

is also divisible by 17 without actually evaluating it.

Because of the result det $(A^T) = $ det A, every property of the determinant which involves the rows of A implies another property of the determinant involving the columns of A. For example, the determinant is linear in each *column*, reverses sign if two *columns* are interchanged, is unaffected if a multiple of one *column* is added to another *column*, and so on.

To begin, multiply the first column of A by 1000, the second column by 100, and the third column by 10. The determinant of the resulting matrix will be $1000 \cdot 100 \cdot 10$ times greater than the determinant of A:

$$\det \begin{bmatrix} 1000 & 500 & 40 & 7 \\ 2000 & 300 & 20 & 9 \\ 3000 & 800 & 90 & 3 \\ 4000 & 400 & 70 & 1 \end{bmatrix} = 1000 \cdot 100 \cdot 10 \cdot \det \begin{bmatrix} 1 & 5 & 4 & 7 \\ 2 & 3 & 2 & 9 \\ 3 & 8 & 9 & 3 \\ 4 & 4 & 7 & 1 \end{bmatrix}$$

Next, add the second, third, and fourth columns of this new matrix to its first column. None of these column operations changes the determinant; thus,

$$10^6 \det \begin{bmatrix} 1 & 5 & 4 & 7 \\ 2 & 3 & 2 & 9 \\ 3 & 8 & 9 & 3 \\ 4 & 4 & 7 & 1 \end{bmatrix} = \det \begin{bmatrix} 1000 & 500 & 40 & 7 \\ 2000 & 300 & 20 & 9 \\ 3000 & 800 & 90 & 3 \\ 4000 & 400 & 70 & 1 \end{bmatrix}$$

$$= \det \begin{bmatrix} 1547 & 500 & 40 & 7 \\ 2329 & 300 & 20 & 9 \\ 3893 & 800 & 90 & 3 \\ 4471 & 400 & 70 & 1 \end{bmatrix}$$

Since each entry in the first column of this latest matrix is divisible by 17, every term in the Laplace expansion by the first column will be divisible by 17, and thus the sum of these terms—which gives the determinant—will be divisible by 17. Since 17 divides $10^6 \det A$, 17 must divide $\det A$ because 17 is prime and doesn't divide 10^6. ∎

Example 22: A useful concept in higher-dimensional calculus (in connection with the change-of-variables formula for multiple integrals, for example) is that of the **Jacobian** of a mapping. Let x and y be given as functions of the independent variables u and v:

$$x = x(u, \ v)$$
$$y = y(u, \ v)$$

The Jacobian of the map $(u, \ v) \mapsto (x, \ y)$, a quantity denoted by the symbol $\partial(x, \ y)/\partial(u, \ v)$, is defined to be the following determinant:

$$\frac{\partial(x, \ y)}{\partial(u, \ v)} = \det \begin{bmatrix} \partial x/\partial u & \partial x/\partial v \\ \partial y/\partial u & \partial y/\partial v \end{bmatrix}$$

To illustrate, consider the *polar coordinate* transformation,

$$x = r\cos\theta \qquad (*)$$
$$y = r\sin\theta$$

The Jacobian of this mapping, $(r, \ \theta) \mapsto (x, \ y)$, is

$$\frac{\partial(x,\,y)}{\partial(r,\,\theta)} = \det \begin{bmatrix} \partial x/\partial r & \partial x/\partial \theta \\ \partial y/\partial r & \partial y/\partial \theta \end{bmatrix}$$

$$= \det \begin{bmatrix} \cos\theta & -r\sin\theta \\ \sin\theta & r\cos\theta \end{bmatrix}$$

$$= r(\cos^2\theta + \sin^2\theta)$$

$$= r$$

The fact that the Jacobian of this transformation is equal to r accounts for the factor of r in the familiar formula

$$\iint_R f(x,\,y)\,dx\,dy = \iint_{R'} f(r\cos\theta,\,r\sin\theta)\,r\,dr\,d\theta$$
$$\uparrow$$

where R' is the region in the r-θ plane mapped by (*) to the region of integration R in the x-y plane.

The Jacobian can also be extended to three variables. For example, a point in 3-space can be specified by giving its *spherical coordinates*—ρ, ϕ, and θ—which are related to the usual rectangular coordinates—x, y, and z—by the equations

$$x = \rho\sin\phi\cos\theta$$
$$y = \rho\sin\phi\sin\theta$$
$$z = \rho\cos\phi$$

See Figure 52.

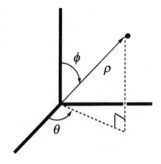

■ Figure 52 ■

The Jacobian of the mapping $(\rho, \phi, \theta) \mapsto (x, y, z)$ is

$$\frac{\partial(x, y, z)}{\partial(\rho, \phi, \theta)} = \det \begin{bmatrix} \partial x/\partial\rho & \partial x/\partial\phi & \partial x/\partial\theta \\ \partial y/\partial\rho & \partial y/\partial\phi & \partial y/\partial\theta \\ \partial z/\partial\rho & \partial z/\partial\phi & \partial z/\partial\theta \end{bmatrix}$$

$$= \det \begin{bmatrix} \sin\phi\cos\theta & \rho\cos\phi\cos\theta & -\rho\sin\phi\sin\theta \\ \sin\phi\sin\theta & \rho\cos\phi\sin\theta & \rho\sin\phi\cos\theta \\ \cos\phi & -\rho\sin\phi & 0 \end{bmatrix}$$

By a Laplace expansion along the third row,

$$\frac{\partial(x, y, z)}{\partial(\rho, \phi, \theta)} = \cos\phi \begin{vmatrix} \rho\cos\phi\cos\theta & -\rho\sin\phi\sin\theta \\ \rho\cos\phi\sin\theta & \rho\sin\phi\cos\theta \end{vmatrix}$$

$$- (-\rho\sin\phi) \begin{vmatrix} \sin\phi\cos\theta & -\rho\sin\phi\sin\theta \\ \sin\phi\sin\theta & \rho\sin\phi\cos\theta \end{vmatrix}$$

$$= \cos\phi \left[\rho^2\cos\phi\sin\phi(\cos^2\theta + \sin^2\theta) \right]$$

$$+ \rho\sin\phi \left[\rho\sin^2\phi(\cos^2\theta + \sin^2\theta) \right]$$

$$= \rho^2\sin\phi\cos^2\phi + \rho^2\sin\phi\sin^2\phi$$

$$= \rho^2\sin\phi$$

The fact that the Jacobian of this transformation is equal to $\rho^2 \sin\phi$ accounts for the factor of $\rho^2 \sin\phi$ in the formula for changing the variables in a triple integral from rectangular to spherical coordinates:

$$\iiint_R f(x,y,z)\, dx\, dy\, dz = \iiint_{R'} f(\rho\sin\phi\cos\theta, \rho\sin\phi\sin\theta, \rho\cos\phi)$$
$$\times \underbrace{\rho^2 \sin\phi}\, d\rho\, d\phi\, d\theta \quad \blacksquare$$

Laplace expansions following row-reduction. The utility of the Laplace expansion method for evaluating a determinant is enhanced when it is preceded by elementary row operations. If such operations are performed on a matrix, the number of zeros in a given column can be increased, thereby decreasing the number of nonzero terms in the Laplace expansion along that column.

Example 23: Evaluate the determinant of the matrix

$$A = \begin{bmatrix} 1 & -2 & 4 \\ 2 & -1 & 1 \\ -3 & 1 & -2 \end{bmatrix}$$

The following row-reduction operations, because they simply involve adding a multiple of one row to another, do not alter the value of the determinant:

$$\begin{bmatrix} 1 & -2 & 4 \\ 2 & -1 & 1 \\ -3 & 1 & -2 \end{bmatrix} \xrightarrow[\begin{array}{c}-2\mathbf{r}_1 \text{ added to } \mathbf{r}_2\\ 3\mathbf{r}_1 \text{ added to } \mathbf{r}_3\end{array}]{} \begin{bmatrix} 1 & -2 & 4 \\ 0 & 3 & -7 \\ 0 & -5 & 10 \end{bmatrix}$$

Now, when the determinant of this latter matrix is computed using the Laplace expansion by the first column, only one nonzero term remains:

$$\begin{vmatrix} 1 & -2 & 4 \\ 0 & 3 & -7 \\ 0 & -5 & 10 \end{vmatrix} = 1 \begin{vmatrix} 3 & -7 \\ -5 & 10 \end{vmatrix} = (3)(10) - (-5)(-7) = -5$$

Therefore, det $A = -5$. ∎

Example 24: Evaluate the determinant of the matrix

$$A = \begin{bmatrix} 5 & 1 & 1 \\ -3 & 3 & -8 \\ 2 & 0 & 4 \end{bmatrix}$$

In order to avoid generating many noninteger entries during the row-reduction process, a factor of 2 is first divided out of the bottom row. Since multiplying a row by a scalar multiplies the determinant by that scalar,

$$\det \begin{bmatrix} 5 & 1 & 1 \\ -3 & 3 & -8 \\ 2 & 0 & 4 \end{bmatrix} = 2 \det \begin{bmatrix} 5 & 1 & 1 \\ -3 & 3 & -8 \\ 1 & 0 & 2 \end{bmatrix}$$

Now, because the elementary row operations

$$\begin{bmatrix} 5 & 1 & 1 \\ -3 & 3 & -8 \\ 1 & 0 & 2 \end{bmatrix} \xrightarrow[\substack{3r_3 \text{ added to } r_2}]{-5r_3 \text{ added to } r_1} \begin{bmatrix} 0 & 1 & -9 \\ 0 & 3 & -2 \\ 1 & 0 & 2 \end{bmatrix}$$

do not change the determinant, Laplace expansion by the first column of this latter matrix completes the evaluation of the determinant of A:

$$\det\begin{bmatrix} 5 & 1 & 1 \\ -3 & 3 & -8 \\ 2 & 0 & 4 \end{bmatrix} = 2\det\begin{bmatrix} 5 & 1 & 1 \\ -3 & 3 & -8 \\ 1 & 0 & 2 \end{bmatrix}$$

$$= 2\det\begin{bmatrix} 0 & 1 & -9 \\ 0 & 3 & -2 \\ 1 & 0 & 2 \end{bmatrix}$$

$$= 2\left(1\cdot\det\begin{bmatrix} 1 & -9 \\ 3 & -2 \end{bmatrix}\right)$$

$$= 2(-2+27)$$

$$= 50 \quad \blacksquare$$

Cramer's Rule

Consider the general 2 by 2 linear system

$$a_{11}x + a_{12}y = b_1$$
$$a_{21}x + a_{22}y = b_2$$

Multiplying the first equation by a_{22}, the second by $-a_{12}$, and adding the results eliminates y and permits evaluation of x:

$$a_{11}a_{22}x + a_{12}a_{22}y = a_{22}b_1$$
$$\underline{-a_{12}a_{21}x - a_{12}a_{22}y = -a_{12}b_2}$$
$$x(a_{11}a_{22} - a_{12}a_{21}) = a_{22}b_1 - a_{12}b_2$$

$$x = \frac{a_{22}b_1 - a_{12}b_2}{a_{11}a_{22} - a_{12}a_{21}}$$

assuming that $a_{11}a_{22} - a_{12}a_{21} \neq 0$. Similarly, multiplying the first equation by $-a_{21}$, the second by a_{11}, and adding the results eliminates x and determines y:

$$-a_{11}a_{21}x - a_{12}a_{21}y = -a_{21}b_1$$
$$\underline{a_{11}a_{21}x + a_{11}a_{22}y = a_{11}b_2}$$
$$y(a_{11}a_{22} - a_{12}a_{21}) = a_{11}b_2 - a_{21}b_1$$

$$y = \frac{a_{11}b_2 - a_{21}b_1}{a_{11}a_{22} - a_{12}a_{21}}$$

again assuming that $a_{11}a_{22} - a_{12}a_{21} \neq 0$. These expressions for x and y can be written in terms of determinants as follows:

$$x = \frac{a_{22}b_1 - a_{12}b_2}{a_{11}a_{22} - a_{12}a_{21}} = \frac{\begin{vmatrix} b_1 & a_{12} \\ b_2 & a_{22} \end{vmatrix}}{\begin{vmatrix} a_{11} & a_{12} \\ a_{21} & a_{22} \end{vmatrix}}$$

and

$$y = \frac{a_{11}b_2 - a_{21}b_1}{a_{11}a_{22} - a_{12}a_{21}} = \frac{\begin{vmatrix} a_{11} & b_1 \\ a_{21} & b_2 \end{vmatrix}}{\begin{vmatrix} a_{11} & a_{12} \\ a_{21} & a_{22} \end{vmatrix}}$$

If the original system is written in matrix form,

$$\begin{bmatrix} a_{11} & a_{12} \\ a_{21} & a_{22} \end{bmatrix} \begin{bmatrix} x \\ y \end{bmatrix} = \begin{bmatrix} b_1 \\ b_2 \end{bmatrix}$$

then the denominators in the above expressions for the unknowns x and y are both equal to the determinant of the coefficient matrix. Furthermore, the numerator in the expression for the first unknown, x, is equal to the determinant of the matrix that results when the first column of the coefficient matrix is replaced by the column of constants, and the numerator in the expression for the second unknown, y, is equal to the determinant of the matrix that results when the second column of the coefficient matrix is replaced by the column of constants. This is **Cramer's Rule** for a 2 by 2 linear system.

Extending the pattern to a 3 by 3 linear system,

$$\begin{bmatrix} a_{11} & a_{12} & a_{13} \\ a_{21} & a_{22} & a_{23} \\ a_{31} & a_{32} & a_{33} \end{bmatrix} \begin{bmatrix} x \\ y \\ z \end{bmatrix} = \begin{bmatrix} b_1 \\ b_2 \\ b_3 \end{bmatrix}$$

Cramer's Rule says that if the determinant of the coefficient matrix is nonzero, then expressions for the unknowns x, y, and z take on the following form:

$$x = \frac{\begin{vmatrix} b_1 & a_{12} & a_{13} \\ b_2 & a_{22} & a_{23} \\ b_3 & a_{32} & a_{33} \end{vmatrix}}{\begin{vmatrix} a_{11} & a_{12} & a_{13} \\ a_{21} & a_{22} & a_{23} \\ a_{31} & a_{32} & a_{33} \end{vmatrix}}, \quad y = \frac{\begin{vmatrix} a_{11} & b_1 & a_{13} \\ a_{21} & b_2 & a_{23} \\ a_{31} & b_3 & a_{33} \end{vmatrix}}{\begin{vmatrix} a_{11} & a_{12} & a_{13} \\ a_{21} & a_{22} & a_{23} \\ a_{31} & a_{32} & a_{33} \end{vmatrix}}, \quad z = \frac{\begin{vmatrix} a_{11} & a_{12} & b_1 \\ a_{21} & a_{22} & b_2 \\ a_{31} & a_{32} & b_3 \end{vmatrix}}{\begin{vmatrix} a_{11} & a_{12} & a_{13} \\ a_{21} & a_{22} & a_{23} \\ a_{31} & a_{32} & a_{33} \end{vmatrix}}$$

The general form of Cramer's Rule reads as follows: A system of n linear equations in n unknowns, written in matrix form $A\mathbf{x} = \mathbf{b}$ as

$$\begin{bmatrix} a_{11} & a_{12} & \cdots & a_{1n} \\ a_{21} & a_{22} & \cdots & a_{2n} \\ \vdots & \vdots & & \vdots \\ a_{n1} & a_{n2} & \cdots & a_{nn} \end{bmatrix} \begin{bmatrix} x_1 \\ x_2 \\ \vdots \\ x_n \end{bmatrix} = \begin{bmatrix} b_1 \\ b_2 \\ \vdots \\ b_n \end{bmatrix}$$

will have a unique solution if det $A \neq 0$, and in this case, the value of the unknown x_j is given by the expression

$$x_j = \frac{\det A_j}{\det A}$$

where A_j is the matrix that results when column j of the coefficient matrix A is replaced by the column matrix $\mathbf{b}$.

Two important theoretical results about square systems follow from Cramer's Rule:

Theorem F. A square system $A\mathbf{x} = \mathbf{b}$ will have a unique solution for every column matrix $\mathbf{b}$ if and only if det $A \neq 0$.

Theorem G. A homogeneous square system $A\mathbf{x} = \mathbf{0}$ will have only the trivial solution $\mathbf{x} = \mathbf{0}$ if and only if det $A \neq 0$.

Although Cramer's Rule is of theoretical importance because it gives a formula for the unknowns, it is generally not an efficient solution method, especially for large systems. Gaussian elimination is still the method of choice. However, Cramer's Rule can be useful when, for example, the value of only one unknown is needed.

Example 25: Use Cramer's Rule to find the value of y given that

$$x + y - 2z = -10$$
$$2x - y + 3z = -1$$
$$4x + 6y + z = 2$$

Since this linear system is equivalent to the matrix equation

$$\begin{bmatrix} 1 & 1 & -2 \\ 2 & -1 & 3 \\ 4 & 6 & 1 \end{bmatrix} \begin{bmatrix} x \\ y \\ z \end{bmatrix} = \begin{bmatrix} -10 \\ -1 \\ 2 \end{bmatrix}$$

Cramer's Rule implies that the second unknown, y, is given by the expression

$$y = \frac{\begin{vmatrix} 1 & -10 & -2 \\ 2 & -1 & 3 \\ 4 & 2 & 1 \end{vmatrix}}{\begin{vmatrix} 1 & 1 & -2 \\ 2 & -1 & 3 \\ 4 & 6 & 1 \end{vmatrix}} \quad (*)$$

assuming that the denominator—the determinant of the coefficient matrix—is not zero. Row-reduction, followed by Laplace expansion along the first column, evaluates these determinants:

$$\begin{vmatrix} 1 & -10 & -2 \\ 2 & -1 & 3 \\ 4 & 2 & 1 \end{vmatrix} \xrightarrow[\substack{-2\mathbf{r}_1 \text{ added to } \mathbf{r}_2 \\ -4\mathbf{r}_1 \text{ added to } \mathbf{r}_3}]{} \begin{vmatrix} 1 & -10 & -2 \\ 0 & 19 & 7 \\ 0 & 42 & 9 \end{vmatrix} = \begin{vmatrix} 19 & 7 \\ 42 & 9 \end{vmatrix}$$

$$= 171 - 294$$
$$= -123$$

$$\begin{vmatrix} 1 & 1 & -2 \\ 2 & -1 & 3 \\ 4 & 6 & 1 \end{vmatrix} \xrightarrow[\substack{-2\mathbf{r}_1 \text{ added to } \mathbf{r}_2 \\ -4\mathbf{r}_1 \text{ added to } \mathbf{r}_3}]{} \begin{vmatrix} 1 & 1 & -2 \\ 0 & -3 & 7 \\ 0 & 2 & 9 \end{vmatrix} = \begin{vmatrix} -3 & 7 \\ 2 & 9 \end{vmatrix}$$

$$= -27 - 14$$
$$= -41$$

With these calculations, (*) implies

$$y = \frac{-123}{-41} = 3 \quad \blacksquare$$

The Classical Adjoint of a Square Matrix

Let $A = [a_{ij}]$ be a square matrix. The transpose of the matrix whose (i, j) entry is the a_{ij} cofactor is called the classical **adjoint** of A:

$$\text{Adj } A = [\text{cof}(a_{ij})]^{\text{T}}$$

Example 26: Find the adjoint of the matrix

$$A = \begin{bmatrix} 1 & -1 & 2 \\ 4 & 0 & 6 \\ 0 & 1 & -1 \end{bmatrix}$$

The first step is to evaluate the cofactor of every entry:

$$\text{cof}(a_{11}) = + \begin{vmatrix} 0 & 6 \\ 1 & -1 \end{vmatrix} = -6 \qquad \text{cof}(a_{12}) = - \begin{vmatrix} 4 & 6 \\ 0 & -1 \end{vmatrix} = 4$$

$$\text{cof}(a_{21}) = - \begin{vmatrix} -1 & 2 \\ 1 & -1 \end{vmatrix} = 1 \qquad \text{cof}(a_{22}) = + \begin{vmatrix} 1 & 2 \\ 0 & -1 \end{vmatrix} = -1$$

$$\text{cof}(a_{31}) = + \begin{vmatrix} -1 & 2 \\ 0 & 6 \end{vmatrix} = -6 \qquad \text{cof}(a_{32}) = - \begin{vmatrix} 1 & 2 \\ 4 & 6 \end{vmatrix} = 2$$

$$\text{cof}(a_{13}) = + \begin{vmatrix} 4 & 0 \\ 0 & 1 \end{vmatrix} = 4$$

$$\text{cof}(a_{23}) = - \begin{vmatrix} 1 & -1 \\ 0 & 1 \end{vmatrix} = -1$$

$$\text{cof}(a_{33}) = + \begin{vmatrix} 1 & -1 \\ 4 & 0 \end{vmatrix} = 4$$

Therefore,

$$\text{Adj } A = [\text{cof}(a_{ij})]^T = \begin{bmatrix} -6 & 4 & 4 \\ 1 & -1 & -1 \\ -6 & 2 & 4 \end{bmatrix}^T = \begin{bmatrix} -6 & 1 & -6 \\ 4 & -1 & 2 \\ 4 & -1 & 4 \end{bmatrix} \qquad \blacksquare$$

Why form the adjoint matrix? First, verify the following calculation where the matrix A above is multiplied by its adjoint:

$$A \cdot \text{Adj } A = \begin{bmatrix} 1 & -1 & 2 \\ 4 & 0 & 6 \\ 0 & 1 & -1 \end{bmatrix} \begin{bmatrix} -6 & 1 & -6 \\ 4 & -1 & 2 \\ 4 & -1 & 4 \end{bmatrix} = \begin{bmatrix} -2 & 0 & 0 \\ 0 & -2 & 0 \\ 0 & 0 & -2 \end{bmatrix} = -2I$$

$$(*)$$

Now, since a Laplace expansion by the first column of A gives

$$\det A = \begin{vmatrix} 1 & -1 & 2 \\ 4 & 0 & 6 \\ 0 & 1 & -1 \end{vmatrix} = 1\begin{vmatrix} 0 & 6 \\ 1 & -1 \end{vmatrix} - 4\begin{vmatrix} -1 & 2 \\ 1 & -1 \end{vmatrix} = -6 - 4(-1) = -2$$

equation (*) becomes

$$A \cdot \text{Adj}\, A = (\det A)I$$

This result gives the following equation for the inverse of A:

$$\boxed{A^{-1} = \frac{\text{Adj}\, A}{\det A}}$$

By generalizing these calculations to an arbitrary n by n matrix, the following theorem can be proved:

Theorem H. A square matrix A is invertible if and only if its determinant is not zero, and its inverse is obtained by multiplying the adjoint of A by $(\det A)^{-1}$. [Note: A matrix whose determinant is 0 is said to be **singular**; therefore, a matrix is invertible if and only if it is nonsingular.]

Example 27: Determine the inverse of the following matrix by first computing its adjoint:

$$A = \begin{bmatrix} 1 & 2 & 3 \\ 4 & 5 & 6 \\ 7 & 8 & 10 \end{bmatrix}$$

First, evaluate the cofactor of each entry in A:

$$\text{cof}(a_{11}) = + \begin{vmatrix} 5 & 6 \\ 8 & 10 \end{vmatrix} = 2 \qquad \text{cof}(a_{12}) = - \begin{vmatrix} 4 & 6 \\ 7 & 10 \end{vmatrix} = 2$$

$$\text{cof}(a_{21}) = - \begin{vmatrix} 2 & 3 \\ 8 & 10 \end{vmatrix} = 4 \qquad \text{cof}(a_{22}) = + \begin{vmatrix} 1 & 3 \\ 7 & 10 \end{vmatrix} = -11$$

$$\text{cof}(a_{31}) = + \begin{vmatrix} 2 & 3 \\ 5 & 6 \end{vmatrix} = -3 \qquad \text{cof}(a_{32}) = - \begin{vmatrix} 1 & 3 \\ 4 & 6 \end{vmatrix} = 6$$

$$\text{cof}(a_{13}) = + \begin{vmatrix} 4 & 5 \\ 7 & 8 \end{vmatrix} = -3$$

$$\text{cof}(a_{23}) = - \begin{vmatrix} 1 & 2 \\ 7 & 8 \end{vmatrix} = 6$$

$$\text{cof}(a_{33}) = + \begin{vmatrix} 1 & 2 \\ 4 & 5 \end{vmatrix} = -3$$

These computations imply that

$$\text{Adj} A = [\text{cof}(a_{ij})]^T = \begin{bmatrix} 2 & 2 & -3 \\ 4 & -11 & 6 \\ -3 & 6 & -3 \end{bmatrix}^T = \begin{bmatrix} 2 & 4 & -3 \\ 2 & -11 & 6 \\ -3 & 6 & -3 \end{bmatrix}$$

Now, since Laplace expansion along the first row gives

$$\det A = \det \begin{bmatrix} 1 & 2 & 3 \\ 4 & 5 & 6 \\ 7 & 8 & 10 \end{bmatrix}$$

$$= 1 \begin{vmatrix} 5 & 6 \\ 8 & 10 \end{vmatrix} - 2 \begin{vmatrix} 4 & 6 \\ 7 & 10 \end{vmatrix} + 3 \begin{vmatrix} 4 & 5 \\ 7 & 8 \end{vmatrix}$$

$$= 2 - 2(-2) + 3(-3)$$

$$= -3$$

the inverse of A is

$$A^{-1} = \frac{\operatorname{Adj} A}{\det A} = -\frac{1}{3}\begin{bmatrix} 2 & 4 & -3 \\ 2 & -11 & 6 \\ -3 & 6 & -3 \end{bmatrix} = \begin{bmatrix} -\frac{2}{3} & -\frac{4}{3} & 1 \\ -\frac{2}{3} & \frac{11}{3} & -2 \\ 1 & -2 & 1 \end{bmatrix}$$

which may be verified by checking that $AA^{-1} = A^{-1}A = I$. ■

Example 28: If A is an invertible n by n matrix, compute the determinant of Adj A in terms of det A.

Because A is invertible, the equation $A^{-1} = \operatorname{Adj} A/\det A$ implies

$$\operatorname{Adj} A = (\det A) \cdot A^{-1}$$

Recall from Example 13 that if B is $n \times n$ and k is a scalar, then $\det(kB) = k^n \det B$. Applying this formula with $k = \det A$ and $B = A^{-1}$ gives

$$\det[(\det A) \cdot A^{-1}] = (\det A)^n \cdot (\det A^{-1})$$
$$= (\det A)^n \cdot (\det A)^{-1} = (\det A)^{n-1}$$

Thus,

$$\det(\operatorname{Adj} A) = (\det A)^{n-1} \quad ■$$

Example 29: Show that the adjoint of the adjoint of A is guaranteed to equal A if A is an invertible 2 by 2 matrix, but not if A is an invertible square matrix of higher order.

First, the equation $A \cdot \mathrm{Adj}\, A = (\det A)I$ can be rewritten

$$\mathrm{Adj}\, A = (\det A) \cdot A^{-1}$$

which implies

$$\mathrm{Adj}(\mathrm{Adj}\, A) = \det(\mathrm{Adj}\, A) \cdot (\mathrm{Adj}\, A)^{-1} \quad (*)$$

Next, the equation $A \cdot \mathrm{Adj}\, A = (\det A)I$ also implies

$$(\mathrm{Adj}\, A)^{-1} = \frac{A}{\det A}$$

This expression, along with the result of Example 28, transforms (*) into

$$\mathrm{Adj}(\mathrm{Adj}\, A) = \det(\mathrm{Adj}\, A) \cdot (\mathrm{Adj}\, A)^{-1}$$
$$= (\det A)^{n-1} \cdot \frac{A}{\det A}$$
$$\mathrm{Adj}(\mathrm{Adj}\, A) = (\det A)^{n-2} A$$

where n is the size of the square matrix A. If $n = 2$, then $(\det A)^{n-2} = (\det A)^0 = 1$—since $\det A \neq 0$—which implies Adj (Adj A) = A, as desired. However, if $n > 2$, then $(\det A)^{n-2}$ will not equal 1 for every nonzero value of det A, so Adj (Adj A) will not necessarily equal A. Yet this proof does show that whatever the size of the matrix, Adj (Adj A) will equal A if det $A = 1$. ■

Example 30: Consider the vector space $C^2(a, b)$ of functions which have a continuous second derivative on the interval $(a, b) \subset \mathbf{R}$. If f, g, and h are functions in this space, then the following determinant,

$$\det \begin{bmatrix} f & g & h \\ f' & g' & h' \\ f'' & g'' & h'' \end{bmatrix}$$

is called the **Wronskian** of f, g, and h. What does the value of the Wronskian say about the linear independence of the functions f, g, and h?

The functions f, g, and h are linearly independent if the only scalars c_1, c_2, and c_3 which satisfy the equation

$$c_1 f + c_2 g + c_3 h = 0 \quad (*)$$

are $c_1 = c_2 = c_3 = 0$. One way to obtain three equations to solve for the three unknowns c_1, c_2, and c_3 is to differentiate (*) and then to differentiate it again. The result is the system

$$c_1 f + c_2 g + c_3 h = 0$$
$$c_1 f' + c_2 g' + c_3 h' = 0$$
$$c_1 f'' + c_2 g'' + c_3 h'' = 0$$

which can be written in matrix form as

$$\begin{bmatrix} f & g & h \\ f' & g' & h' \\ f'' & g'' & h'' \end{bmatrix} \mathbf{c} = \mathbf{0} \quad (**)$$

where $\mathbf{c} = (c_1, c_2, c_3)^{\mathrm{T}}$. A homogeneous square system—such as this one—has only the trivial solution if and only if the determinant of the coefficient matrix is nonzero. But if $\mathbf{c} = \mathbf{0}$ is the only solution to (**), then $c_1 = c_2 = c_3 = 0$ is the only solution to (*), and the functions f, g, and h are linearly independent. Therefore,

f, g, and h are linearly independent $\Leftrightarrow \det \begin{bmatrix} f & g & h \\ f' & g' & h' \\ f'' & g'' & h'' \end{bmatrix} \neq 0$

To illustrate this result, consider the functions f, g, and h defined by the equations

$$f(x) = \sin^2 x, \quad g(x) = \cos^2 x, \quad \text{and} \quad h(x) \equiv 3$$

Since the Wronskian of these functions is

$$\det \begin{bmatrix} \sin^2 x & \cos^2 x & 3 \\ \sin 2x & -\sin 2x & 0 \\ 2\cos 2x & -2\cos 2x & 0 \end{bmatrix} = 3 \begin{vmatrix} \sin 2x & -\sin 2x \\ 2\cos 2x & -2\cos 2x \end{vmatrix}$$

$$= 3(-2\sin 2x \cos 2x + 2\sin 2x \cos 2x)$$

$$\equiv 0$$

these functions are linearly dependent. This same result was demonstrated in Example 56, page 195 by writing the following nontrivial linear combination of these functions that gave zero: $3f + 3g - h \equiv 0$.

Here's another illustration. Consider the functions f, g, and h in the space $C^2(1/2, \infty)$ defined by the equations

$$f(x) = e^x, \quad g(x) = x, \quad \text{and} \quad h(x) = \ln x$$

By a Laplace expansion along the second column, the Wronskian of these functions is

$$W(x) = \det \begin{bmatrix} e^x & x & \ln x \\ e^x & 1 & 1/x \\ e^x & 0 & -1/x^2 \end{bmatrix} = -x \begin{vmatrix} e^x & 1/x \\ e^x & -1/x^2 \end{vmatrix} + \begin{vmatrix} e^x & \ln x \\ e^x & -1/x^2 \end{vmatrix}$$

$$= e^x \left(\frac{1}{x} + 1 - \frac{1}{x^2} - \ln x \right)$$

Since this function is not identically zero on the interval (1/2, ∞)—for example, when $x = 1$, $W(x) = W(1) = e \neq 0$—the functions f, g, and h are linearly independent. ∎

Given two sets A and B, a function f which accepts as input an element in A and produces as output an element in B is written $f : A \to B$. This is read "f maps A into B," and the words *map* and *function* are used interchangeably. The statement "f maps the element a in A to the element b in B" is symbolized $f(a) = b$ and read "f of a equals b." The set A is called the **domain** of f, and b is the **image** of a by f. The purpose here is to study a certain class of functions between finite-dimensional vector spaces. Throughout this section, vectors will be written as column matrices.

Definition of a Linear Transformation

Let V and W be vector spaces. A function $f: V \to W$ is said to be a **linear map** or **linear transformation** if both of the following conditions hold for all vectors $\mathbf{x}$ and $\mathbf{y}$ in V and any scalar k:

$$f(\mathbf{x} + \mathbf{y}) = f(\mathbf{x}) + f(\mathbf{y})$$
$$f(k\mathbf{x}) = k f(\mathbf{x})$$

These equations express the fact that a map is linear if and only if it is compatible with the vector space operations of vector addition and scalar multiplication.

Example 1: Consider the function $f: \mathbf{R}^2 \to \mathbf{R}^3$ defined by the equation

$$f\begin{bmatrix} x_1 \\ x_2 \end{bmatrix} = \begin{bmatrix} x_1 \\ x_1 + x_2 \\ 2x_1 - 3x_2 \end{bmatrix}$$

Since

$$f(\mathbf{x} + \mathbf{y}) = f\left(\begin{bmatrix} x_1 \\ x_2 \end{bmatrix} + \begin{bmatrix} y_1 \\ y_2 \end{bmatrix} \right)$$

$$= f\begin{bmatrix} x_1 + y_1 \\ x_2 + y_2 \end{bmatrix}$$

$$= \begin{bmatrix} x_1 + y_1 \\ x_1 + y_1 + x_2 + y_2 \\ 2(x_1 + y_1) - 3(x_2 + y_2) \end{bmatrix}$$

$$= \begin{bmatrix} x_1 \\ x_1 + x_2 \\ 2x_1 - 3x_2 \end{bmatrix} + \begin{bmatrix} y_1 \\ y_1 + y_2 \\ 2y_1 - 3y_2 \end{bmatrix}$$

$$= f\begin{bmatrix} x_1 \\ x_2 \end{bmatrix} + f\begin{bmatrix} y_1 \\ y_2 \end{bmatrix}$$

$$= f(\mathbf{x}) + f(\mathbf{y})$$

and

$$f(k\mathbf{x}) = f\left(k\begin{bmatrix} x_1 \\ x_2 \end{bmatrix}\right)$$

$$= f\begin{bmatrix} kx_1 \\ kx_2 \end{bmatrix}$$

$$= \begin{bmatrix} kx_1 \\ kx_1 + kx_2 \\ 2(kx_1) - 3(kx_2) \end{bmatrix}$$

$$= k\begin{bmatrix} x_1 \\ x_1 + x_2 \\ 2x_1 - 3x_2 \end{bmatrix}$$

$$= kf(\mathbf{x})$$

the two conditions for linearity are satisfied; thus, f is a linear transformation. ∎

Example 2: Consider the function $g: \mathbf{R}^2 \rightarrow \mathbf{R}^3$ defined by the equation

$$g\begin{bmatrix} x_1 \\ x_2 \end{bmatrix} = \begin{bmatrix} x_1 \\ x_1 + x_2 \\ 2x_1 - 3x_2 + 1 \end{bmatrix}$$

This function is not linear. For example, if $k = 2$ and $\mathbf{x} = (1, 1)^T$, then

$$g(k\mathbf{x}) = g\left(2\begin{bmatrix} 1 \\ 1 \end{bmatrix}\right) = g\begin{bmatrix} 2 \\ 2 \end{bmatrix} = \begin{bmatrix} 2 \\ 4 \\ -1 \end{bmatrix}$$

but

$$kg(\mathbf{x}) = 2g\begin{bmatrix} 1 \\ 1 \end{bmatrix} = 2\begin{bmatrix} 1 \\ 2 \\ 0 \end{bmatrix} = \begin{bmatrix} 2 \\ 4 \\ 0 \end{bmatrix}$$

Thus, $g(k\mathbf{x}) \neq kg(\mathbf{x})$; this map is not compatible with scalar multiplication. For this particular function, nonlinearity could also have been established by showing that f is incompatible with vector addition. Let $\mathbf{y} = (2, 2)^\mathrm{T}$; then

$$g(\mathbf{x}+\mathbf{y}) = g\left(\begin{bmatrix} 1 \\ 1 \end{bmatrix} + \begin{bmatrix} 2 \\ 2 \end{bmatrix}\right) = g\begin{bmatrix} 3 \\ 3 \end{bmatrix} = \begin{bmatrix} 3 \\ 6 \\ -2 \end{bmatrix}$$

but

$$g(\mathbf{x})+g(\mathbf{y}) = g\begin{bmatrix} 1 \\ 1 \end{bmatrix} + g\begin{bmatrix} 2 \\ 2 \end{bmatrix} = \begin{bmatrix} 1 \\ 2 \\ 0 \end{bmatrix} + \begin{bmatrix} 2 \\ 4 \\ -1 \end{bmatrix} = \begin{bmatrix} 3 \\ 6 \\ -1 \end{bmatrix}$$

The fact that $g(\mathbf{x}+\mathbf{y}) \neq g(\mathbf{x})+g(\mathbf{y})$ also establishes that the function g is not linear. ■

Example 3: If $f: V \to W$ is linear, then $f(\mathbf{0}) = \mathbf{0}$; that is, *a linear transformation always maps the zero vector (in V) to the zero vector (in W)*. One way to prove this is to use the fact that f must be compatible with vector addition. Since $\mathbf{0} = \mathbf{0} + \mathbf{0}$,

$$f(\mathbf{0}) = f(\mathbf{0}+\mathbf{0}) = f(\mathbf{0})+f(\mathbf{0}) = 2f(\mathbf{0})$$

Now, subtracting $f(\mathbf{0})$ from both sides of the equation $f(\mathbf{0}) = 2f(\mathbf{0})$ yields $f(\mathbf{0}) = \mathbf{0}$, as desired.

This observation provides a quick way to determine that a given function is not linear: *If f does not map the zero vector to the zero vector, then f cannot be linear.* To illustrate, the function $g: \mathbf{R}^2 \to \mathbf{R}^3$ defined in Example 2 above could have been proven to be nonlinear by simply noting that g maps the zero vector $\mathbf{0} = (0, 0)^T$ in $\mathbf{R}^2$ to the *nonzero* vector $(0, 0, 1)^T$ in $\mathbf{R}^3$. ∎

Example 4: Let $h: \mathbf{R}^2 \to \mathbf{R}^2$ be defined by the equation $h(x_1, x_2)^T = (x_1 - x_2, x_1 x_2)^T$. Is h linear?

First, check whether the function maps $\mathbf{0}$ to $\mathbf{0}$; if the answer is *no*, then no further analysis is necessary; the function is nonlinear by the result proved in the preceding example. Although the answer is *yes* in this case—h does map $\mathbf{0}$ to $\mathbf{0}$—this does not mean that the function is linear. The guarantee is that if the function does *not* map the zero vector to the zero vector then it is *not* linear; the condition that $\mathbf{0}$ be mapped to $\mathbf{0}$ is necessary, *but not sufficient*, to establish linearity. In fact, h is not linear, as seen by the following calculation: If $\mathbf{x} = (1, 1)^T$ and $k = 2$, then

$$h(k\mathbf{x}) = h\left(2\begin{bmatrix}1\\1\end{bmatrix}\right) = h\begin{bmatrix}2\\2\end{bmatrix} = \begin{bmatrix}0\\4\end{bmatrix}$$

But

$$kh(\mathbf{x}) = 2h\begin{bmatrix}1\\1\end{bmatrix} = 2\begin{bmatrix}0\\1\end{bmatrix} = \begin{bmatrix}0\\2\end{bmatrix}$$

Thus, $h(k\mathbf{x}) \neq kh(\mathbf{x})$. Incompatibility with scalar multiplication proves that h is nonlinear. ■

Example 5: Consider the matrix

$$A = \begin{bmatrix} 1 & 0 \\ 1 & 1 \\ 2 & -3 \end{bmatrix}$$

and define a function $T: \mathbf{R}^2 \to \mathbf{R}^3$ by the equation $T(\mathbf{x}) = A\mathbf{x}$. Is T linear?

Clearly, $T(\mathbf{0}) = A \cdot \mathbf{0} = \mathbf{0}$, so you cannot immediately say that T is nonlinear. In fact, checking the two conditions reveals that T actually is a linear transformation:

$$
\begin{aligned}
T(\mathbf{x} + \mathbf{y}) &= A(\mathbf{x} + \mathbf{y}) \\
&= \begin{bmatrix} 1 & 0 \\ 1 & 1 \\ 2 & -3 \end{bmatrix} \left(\begin{bmatrix} x_1 \\ x_2 \end{bmatrix} + \begin{bmatrix} y_1 \\ y_2 \end{bmatrix} \right) \\
&= \begin{bmatrix} 1 & 0 \\ 1 & 1 \\ 2 & -3 \end{bmatrix} \begin{bmatrix} x_1 + y_1 \\ x_2 + y_2 \end{bmatrix} \\
&= \begin{bmatrix} x_1 + y_1 \\ x_1 + y_1 + x_2 + y_2 \\ 2(x_1 + y_1) - 3(x_2 + y_2) \end{bmatrix}
\end{aligned}
$$

$$= \begin{bmatrix} x_1 \\ x_1 + x_2 \\ 2x_1 - 3x_2 \end{bmatrix} + \begin{bmatrix} y_1 \\ y_1 + y_2 \\ 2y_1 - 3y_2 \end{bmatrix}$$

$$= \begin{bmatrix} 1 & 0 \\ 1 & 1 \\ 2 & -3 \end{bmatrix} \begin{bmatrix} x_1 \\ x_2 \end{bmatrix} + \begin{bmatrix} 1 & 0 \\ 1 & 1 \\ 2 & -3 \end{bmatrix} \begin{bmatrix} y_1 \\ y_2 \end{bmatrix}$$

$$= A\mathbf{x} + A\mathbf{y}$$

$$= T(\mathbf{x}) + T(\mathbf{y})$$

and

$$T(k\mathbf{x}) = A(k\mathbf{x}) = \begin{bmatrix} 1 & 0 \\ 1 & 1 \\ 2 & -3 \end{bmatrix} \left(k \begin{bmatrix} x_1 \\ x_2 \end{bmatrix} \right)$$

$$= \begin{bmatrix} 1 & 0 \\ 1 & 1 \\ 2 & -3 \end{bmatrix} \begin{bmatrix} kx_1 \\ kx_2 \end{bmatrix}$$

$$= \begin{bmatrix} kx_1 \\ kx_1 + kx_2 \\ 2(kx_1) - 3(kx_2) \end{bmatrix}$$

$$= k \begin{bmatrix} x_1 \\ x_1 + x_2 \\ 2x_1 - 3x_2 \end{bmatrix}$$

$$= k \begin{bmatrix} 1 & 0 \\ 1 & 1 \\ 2 & -3 \end{bmatrix} \begin{bmatrix} x_1 \\ x_2 \end{bmatrix}$$

$$= k(A\mathbf{x})$$

$$= kT(\mathbf{x})$$

Since T is compatible with both vector addition and scalar multiplication, T is linear. ∎

Example 6: Consider the polynomial space P_3 and define a map $D: P_3 \to P_2$ by the equation $D(\mathbf{p}) = \mathbf{p}'$. Explicitly, $D(a_0 + a_1 x + a_2 x^2 + a_3 x^3) = a_1 + 2a_2 x + 3a_3 x^2$; this is the differentiation map. Is D linear?

Because the derivative of a sum is equal to the sum of the derivatives,

$$D(\mathbf{p} + \mathbf{q}) = (\mathbf{p} + \mathbf{q})' = \mathbf{p}' + \mathbf{q}' = D(\mathbf{p}) + D(\mathbf{q})$$

and because the derivative of a scalar times a polynomial is equal to the scalar times the derivative of the polynomial,

$$D(k\mathbf{p}) = (k\mathbf{p})' = k\mathbf{p}' = kD(\mathbf{p})$$

the map D is indeed linear. ∎

Example 7: Is the map $T: M_{2\times3} \to M_{3\times2}$ given by $T(A) = A^{\mathrm{T}}$ a linear transformation?

The process of taking the transpose is compatible with addition,

$$T(A + B) = (A + B)^{\mathrm{T}} = A^{\mathrm{T}} + B^{\mathrm{T}} = T(A) + T(B)$$

and with scalar multiplication,

$$T(kA) = (kA)^{\mathrm{T}} = kA^{\mathrm{T}} = kT(A)$$

Thus, T is linear. ■

Example 8: Is the determinant function on 2 by 2 matrices, det: $M_{2 \times 2} \to \mathbf{R}$, a linear transformation?

No. The determinant function is incompatible with addition, as illustrated by the following example. If

$$A = \begin{bmatrix} 1 & 0 \\ 0 & 0 \end{bmatrix} \quad \text{and} \quad B = \begin{bmatrix} 0 & 0 \\ 0 & 1 \end{bmatrix}$$

then

$$\det(A + B) = \det \begin{bmatrix} 1 & 0 \\ 0 & 1 \end{bmatrix} = 1$$

but

$$\det A + \det B = \det \begin{bmatrix} 1 & 0 \\ 0 & 0 \end{bmatrix} + \det \begin{bmatrix} 0 & 0 \\ 0 & 1 \end{bmatrix} = 0 + 0 = 0$$

Thus, $\det(A + B) \neq \det A + \det B$, so det is not a linear map. You may also show that det is nonlinear by providing a counterexample to the statement $\det(kA) = k \det A$. ■

Linear Transformations and Basis Vectors

Theorem I. Let $T: V \to W$ be a linear transformation between the finite-dimensional vector spaces V and W. Then the image of *every* vector in V is completely specified—and, therefore, the action of T is determined—once the images of the vectors in a basis for V are known.

Proof. Let V be n-dimensional with basis $B = \{ \mathbf{v}_1, \mathbf{v}_2, \ldots, \mathbf{v}_n \}$ and assume that the images of these basis vectors, $T(\mathbf{v}_1) = \mathbf{w}_1$, $T(\mathbf{v}_2) = \mathbf{w}_2, \ldots, T(\mathbf{v}_n) = \mathbf{w}_n$ are known. Recall that if $\mathbf{v}$ is an arbitrary vector in V, then $\mathbf{v}$ can be written *uniquely* in terms of the basis vectors; for example,

$$\mathbf{v} = k_1 \mathbf{v}_1 + k_2 \mathbf{v}_2 + \cdots + k_n \mathbf{v}_n$$

The calculation

$$
\begin{aligned}
T(\mathbf{v}) &= T(k_1 \mathbf{v}_1 + k_2 \mathbf{v}_2 + \cdots + k_n \mathbf{v}_n) \\
&= T(k_1 \mathbf{v}_1) + T(k_2 \mathbf{v}_2) + \cdots + T(k_n \mathbf{v}_n) \\
&= k_1 T(\mathbf{v}_1) + k_2 T(\mathbf{v}_2) + \cdots + k_n T(\mathbf{v}_n) \\
&= k_1 \mathbf{w}_1 + k_2 \mathbf{w}_2 + \cdots + k_n \mathbf{w}_n
\end{aligned}
$$

where the first two steps follow from the linearity of T, shows that $T(\mathbf{v})$ is uniquely determined. In short, once you know what T does to the basis vectors, you know what it does to *every* vector. ■

Example 9: Let $T: \mathbf{R}^2 \to \mathbf{R}^2$ be the linear transformation that maps $(1, 0)^T$ to $(2, -3)^T$ and $(0, 1)^T$ to $(-1, 4)^T$. Where does T map the vector $(3, 5)^T$?

Since $(1, 0)^T$ and $(0, 1)^T$ form a basis for $\mathbf{R}^2$, and the images of these vectors are known, the image of every vector in $\mathbf{R}^2$ can be determined. In particular,

$$T\begin{bmatrix} 3 \\ 5 \end{bmatrix} = T\left(3\begin{bmatrix} 1 \\ 0 \end{bmatrix} + 5\begin{bmatrix} 0 \\ 1 \end{bmatrix} \right)$$

$$= 3\,T\begin{bmatrix} 1 \\ 0 \end{bmatrix} + 5\,T\begin{bmatrix} 0 \\ 1 \end{bmatrix}$$

$$= 3\begin{bmatrix} 2 \\ -3 \end{bmatrix} + 5\begin{bmatrix} -1 \\ 4 \end{bmatrix}$$

$$= \begin{bmatrix} 6 \\ -9 \end{bmatrix} + \begin{bmatrix} -5 \\ 20 \end{bmatrix}$$

$$= \begin{bmatrix} 1 \\ 11 \end{bmatrix} \qquad \blacksquare$$

Example 10: Let $T: \mathbf{R}^2 \to \mathbf{R}^2$ be the linear transformation that maps $(1, 1)^T$ to itself and $(1, -3)^T$ to $(5, -15)^T$. Where does T map the vector $(3, 5)^T$?

First, note that the vectors $\mathbf{v}_1 = (1, 1)^T$ and $\mathbf{v}_2 = (1, -3)^T$ form a basis for $\mathbf{R}^2$, since they are linearly independent and span all of $\mathbf{R}^2$. In order to use the information about the images of these basis vectors to compute the image of $\mathbf{v} = (3, 5)^T$, the components of $\mathbf{v}$ relative to this basis must first be determined. The coefficients k_1 and k_2 that satisfy $k_1\mathbf{v}_1 + k_2\mathbf{v}_2 = \mathbf{v}$ are evaluated as follows. The equation

$$k_1\begin{bmatrix}1\\1\end{bmatrix} + k_2\begin{bmatrix}1\\-3\end{bmatrix} = \begin{bmatrix}3\\5\end{bmatrix}$$

$$\begin{bmatrix}1 & 1\\1 & -3\end{bmatrix}\begin{bmatrix}k_1\\k_2\end{bmatrix} = \begin{bmatrix}3\\5\end{bmatrix}$$

leads to the augmented matrix

$$\begin{bmatrix}1 & 1 & 3\\1 & -3 & 5\end{bmatrix} \xrightarrow{\;-r_1 \text{ added to } r_2\;} \begin{bmatrix}1 & 1 & 3\\0 & -4 & 2\end{bmatrix}$$

from which it is easy to see that $k_2 = -1/2$ and $k_1 = 7/2$. Therefore,

$$\mathbf{v} = \tfrac{7}{2}\mathbf{v}_1 - \tfrac{1}{2}\mathbf{v}_2$$

so

$$\begin{aligned}T(\mathbf{v}) &= T(\tfrac{7}{2}\mathbf{v}_1 - \tfrac{1}{2}\mathbf{v}_2)\\ &= \tfrac{7}{2}T(\mathbf{v}_1) - \tfrac{1}{2}T(\mathbf{v}_2)\\ &= \tfrac{7}{2}\begin{bmatrix}1\\1\end{bmatrix} - \tfrac{1}{2}\begin{bmatrix}5\\-15\end{bmatrix}\end{aligned}$$

$$T\begin{bmatrix}3\\5\end{bmatrix} = \begin{bmatrix}1\\11\end{bmatrix}$$

The transformation here is, in fact, the same as the one in Example 9. ∎

The Standard Matrix of a Linear Transformation

The details of the verification of the conditions for linearity in Example 5 above were unnecessary. If A is *any* m by n matrix, and $\mathbf{x}$ and $\mathbf{y}$ are n-vectors (written as column matrices), then $A(\mathbf{x} + \mathbf{y}) = A\mathbf{x} + A\mathbf{y}$, and for any scalar k, $A(k\mathbf{x}) = k(A\mathbf{x})$. Therefore, if $T\colon \mathbf{R}^n \to \mathbf{R}^m$ is a map defined by the equation $T(\mathbf{x}) = A\mathbf{x}$, where A is an m by n matrix—that is, if T is given by multiplication by an m by n matrix—then T is *automatically* a linear transformation. Note in particular that the matrix A given in Example 5 generates the function described in Example 1. This is an illustration of the following general result:

Theorem J. Every linear transformation from $\mathbf{R}^n$ to $\mathbf{R}^m$ is given by multiplication by some m by n matrix.

This result raises this question: Given a linear transformation $T\colon \mathbf{R}^n \to \mathbf{R}^m$, how is a **representative matrix** for T computed? That is, how do you find a matrix A such that $T(\mathbf{x}) = A\mathbf{x}$ for every $\mathbf{x}$ in $\mathbf{R}^n$?

The solution begins by recalling the observation that the action of a linear transformation is completely specified once its action on the basis vectors of the domain space is known. Let $T\colon \mathbf{R}^n \to \mathbf{R}^m$ be a given linear map and assume that both $\mathbf{R}^n$ and $\mathbf{R}^m$ are being considered with their *standard* bases, $B = \{\mathbf{e}_1, \mathbf{e}_2, \ldots, \mathbf{e}_n\} \subset \mathbf{R}^n$ and $B' = \{\mathbf{e}'_1, \mathbf{e}'_2, \ldots, \mathbf{e}'_m\} \subset \mathbf{R}^m$. The goal is to find a matrix A—which must necessarily be m by n—such that $T(\mathbf{x}) = A\mathbf{x}$ for every $\mathbf{x}$ in $\mathbf{R}^n$. In particular, $T(\mathbf{x})$ must equal $A\mathbf{x}$ for every $\mathbf{x}$ in the basis B for $\mathbf{R}^n$. Since

$$T(\mathbf{e}_1) = T\begin{bmatrix} 1 \\ 0 \\ \vdots \\ 0 \end{bmatrix} = A\begin{bmatrix} 1 \\ 0 \\ \vdots \\ 0 \end{bmatrix} = \begin{bmatrix} a_{11} & a_{12} & \cdots & a_{1n} \\ a_{21} & a_{22} & \cdots & a_{2n} \\ \vdots & \vdots & & \vdots \\ a_{m1} & a_{m2} & \cdots & a_{mn} \end{bmatrix}\begin{bmatrix} 1 \\ 0 \\ \vdots \\ 0 \end{bmatrix} = \begin{bmatrix} a_{11} \\ a_{21} \\ \vdots \\ a_{m1} \end{bmatrix}$$

$$T(\mathbf{e}_2) = T\begin{bmatrix} 0 \\ 1 \\ \vdots \\ 0 \end{bmatrix} = A\begin{bmatrix} 0 \\ 1 \\ \vdots \\ 0 \end{bmatrix} = \begin{bmatrix} a_{11} & a_{12} & \cdots & a_{1n} \\ a_{21} & a_{22} & \cdots & a_{2n} \\ \vdots & \vdots & & \vdots \\ a_{m1} & a_{m2} & \cdots & a_{mn} \end{bmatrix}\begin{bmatrix} 0 \\ 1 \\ \vdots \\ 0 \end{bmatrix} = \begin{bmatrix} a_{12} \\ a_{22} \\ \vdots \\ a_{m2} \end{bmatrix}$$

$$\vdots$$

$$T(\mathbf{e}_n) = T\begin{bmatrix} 0 \\ 0 \\ \vdots \\ 1 \end{bmatrix} = A\begin{bmatrix} 0 \\ 0 \\ \vdots \\ 1 \end{bmatrix} = \begin{bmatrix} a_{11} & a_{12} & \cdots & a_{1n} \\ a_{21} & a_{22} & \cdots & a_{2n} \\ \vdots & \vdots & & \vdots \\ a_{m1} & a_{m2} & \cdots & a_{mn} \end{bmatrix}\begin{bmatrix} 0 \\ 0 \\ \vdots \\ 1 \end{bmatrix} = \begin{bmatrix} a_{1n} \\ a_{2n} \\ \vdots \\ a_{mn} \end{bmatrix}$$

it is clear that *the images of the basis vectors are the columns of A:*

$$A = \begin{bmatrix} | & | & & | \\ T(\mathbf{e}_1) & T(\mathbf{e}_2) & \cdots & T(\mathbf{e}_n) \\ | & | & & | \end{bmatrix}$$

When $\mathbf{R}^n$ and $\mathbf{R}^m$ are equipped with their standard bases, the matrix representative, A, for a linear transformation $T: \mathbf{R}^n \rightarrow \mathbf{R}^m$ is called the **standard matrix** for T; this is symbolized $A = [T]$.

Example 11: A linear map that maps a vector space into itself is called a **linear operator**. Let $T: \mathbf{R}^2 \rightarrow \mathbf{R}^2$ be the linear operator—given in Example 9—that maps $(1, 0)^T$ to $(2, -3)^T$ and $(0, 1)^T$ to $(-1, 4)^T$. Verify the calculation $T(\mathbf{v}) = (1, 11)^T$, where $\mathbf{v} = (3, 5)^T$, by first computing the standard matrix for T.

The images of the standard basis vectors $(1, 0)^T$ and $(0, 1)^T$ form the columns of the standard matrix. Thus,

$$[T] = \begin{bmatrix} | & | \\ T(\mathbf{e}_1) & T(\mathbf{e}_2) \\ | & | \end{bmatrix} = \begin{bmatrix} 2 & -1 \\ -3 & 4 \end{bmatrix}$$

from which it follows

$$T(\mathbf{v}) = [T]\begin{bmatrix} 3 \\ 5 \end{bmatrix} = \begin{bmatrix} 2 & -1 \\ -3 & 4 \end{bmatrix}\begin{bmatrix} 3 \\ 5 \end{bmatrix} = \begin{bmatrix} 1 \\ 11 \end{bmatrix}$$

as calculated in Example 9. ■

Example 12: Let $T: \mathbf{R}^3 \rightarrow \mathbf{R}^4$ be the linear map defined by the equation

$$T\begin{bmatrix} x_1 \\ x_2 \\ x_3 \end{bmatrix} = \begin{bmatrix} x_1 + 2x_2 + 3x_3 \\ -x_1 + x_2 \\ 4x_2 - x_3 \\ 2x_1 + x_2 - 2x_3 \end{bmatrix}$$

Find the standard matrix for T and use it to compute the image of the vector $\mathbf{v} = (1, 2, 3)^T$.

The images of the standard basis vectors are the columns of the standard matrix. Since

$$T\begin{bmatrix} 1 \\ 0 \\ 0 \end{bmatrix} = \begin{bmatrix} 1 \\ -1 \\ 0 \\ 2 \end{bmatrix}, \quad T\begin{bmatrix} 0 \\ 1 \\ 0 \end{bmatrix} = \begin{bmatrix} 2 \\ 1 \\ 4 \\ 1 \end{bmatrix}, \quad \text{and} \quad T\begin{bmatrix} 0 \\ 0 \\ 1 \end{bmatrix} = \begin{bmatrix} 3 \\ 0 \\ -1 \\ -2 \end{bmatrix}$$

the standard matrix for T is

$$[T] = \begin{bmatrix} 1 & 2 & 3 \\ -1 & 1 & 0 \\ 0 & 4 & -1 \\ 2 & 1 & -2 \end{bmatrix}$$

Since the standard matrix is constructed so that $T(\mathbf{v}) = [T]\mathbf{v}$, the image of $\mathbf{v} = (1, 2, 3)^\mathrm{T}$ is given by the matrix product

$$T(\mathbf{v}) = [T]\mathbf{v} = \begin{bmatrix} 1 & 2 & 3 \\ -1 & 1 & 0 \\ 0 & 4 & -1 \\ 2 & 1 & -2 \end{bmatrix} \begin{bmatrix} 1 \\ 2 \\ 3 \end{bmatrix} = \begin{bmatrix} 14 \\ 1 \\ 5 \\ -2 \end{bmatrix} \quad \blacksquare$$

Example 13: Fix a vector $\mathbf{v}_0 = (a, b, c)^\mathrm{T}$ in $\mathbf{R}^3$ and define a linear operator $T: \mathbf{R}^3 \to \mathbf{R}^3$ by the formula $T(\mathbf{x}) = \mathbf{v}_0 \times \mathbf{x}$. Find the standard matrix for T.

The columns of the standard matrix are the images of the standard basis vectors $\mathbf{e}_1 = \mathbf{i}$, $\mathbf{e}_2 = \mathbf{j}$, and $\mathbf{e}_3 = \mathbf{k}$ under T. Since

$$T(\mathbf{e}_1) = T(\mathbf{i}) = \mathbf{v}_0 \times \mathbf{i} = \begin{vmatrix} \mathbf{i} & \mathbf{j} & \mathbf{k} \\ a & b & c \\ 1 & 0 & 0 \end{vmatrix} = c\mathbf{j} - b\mathbf{k}$$

$$T(\mathbf{e}_2) = T(\mathbf{j}) = \mathbf{v}_0 \times \mathbf{j} = \begin{vmatrix} \mathbf{i} & \mathbf{j} & \mathbf{k} \\ a & b & c \\ 0 & 1 & 0 \end{vmatrix} = -c\mathbf{i} + a\mathbf{k}$$

$$T(\mathbf{e}_3) = T(\mathbf{k}) = \mathbf{v}_0 \times \mathbf{k} = \begin{vmatrix} \mathbf{i} & \mathbf{j} & \mathbf{k} \\ a & b & c \\ 0 & 0 & 1 \end{vmatrix} = b\mathbf{i} - a\mathbf{j}$$

the standard matrix for T is

$$[T] = \begin{bmatrix} | & | & | \\ T(\mathbf{e}_1) & T(\mathbf{e}_2) & T(\mathbf{e}_3) \\ | & | & | \end{bmatrix} = \begin{bmatrix} 0 & -c & b \\ c & 0 & -a \\ -b & a & 0 \end{bmatrix}$$

Example 14: Consider the linear operator T_θ or
by $T_\theta(\mathbf{x}) = A_\theta\mathbf{x}$ where

$$A_\theta = \begin{bmatrix} \cos\theta & -\sin\theta \\ \sin\theta & \cos\theta \end{bmatrix}$$

What does T_θ do to the standard basis vec
$T_\theta(1, 1)^T$ for $\theta = 60°$.

Since T is already a **matrix transf**
ear map of the form $T(\mathbf{x}) = A\mathbf{x}$—the
itself. Since $[T_\theta] = A_\theta$, and the im
vectors are the columns of the sta

$$T(\mathbf{e}_1) = \text{column 1 of } A_\theta = \begin{bmatrix} \cos\theta \\ \sin\theta \end{bmatrix}$$

$$T(\mathbf{e}_2) = \text{column 2 of } A_\theta = \begin{bmatrix} -\sin\theta \\ \cos\theta \end{bmatrix}$$

Figure 53 shows that the effect of this map is to rotate the basis vectors through the angle θ; this transformation is therefore called a **rotation**.

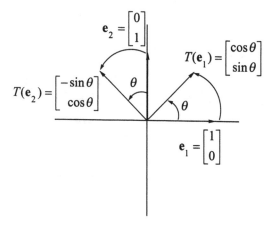

■   Figure 53   ■

$= 60°$, then

$$A_\theta = \begin{bmatrix} \cos\theta & -\sin\theta \\ \sin\theta & \cos\theta \end{bmatrix} = \begin{bmatrix} 1/2 & -\sqrt{3}/2 \\ \sqrt{3}/2 & 1/2 \end{bmatrix}$$

age of the vector $\mathbf{v} = (1, 1)^T$ is

$$\mathbf{v}) = A_\theta \mathbf{v} = \begin{bmatrix} 1/2 & -\sqrt{3}/2 \\ \sqrt{3}/2 & 1/2 \end{bmatrix} \begin{bmatrix} 1 \\ 1 \end{bmatrix} = \frac{1}{2} \begin{bmatrix} 1-\sqrt{3} \\ 1+\sqrt{3} \end{bmatrix}$$

a result that can be verified geometrically. ■

Example 15: Consider the differentiation map $D: P_3 \rightarrow P_2$ given in Example 6 above. Determine the representative matrix for D relative to the standard bases for P_3 and P_2.

Because P_3 is isomorphic to $\mathbf{R}^4$ and P_2 is isomorphic to $\mathbf{R}^3$, the map $D: P_3 \rightarrow P_2$ can be regarded as a map $\tilde{D}: \mathbf{R}^4 \rightarrow \mathbf{R}^3$; furthermore, the matrix for D relative to the standard bases for P_3 and P_2 is the same as the standard matrix for $\tilde{D}: \mathbf{R}^4 \rightarrow \mathbf{R}^3$. Since

$$D(a_0 + a_1 x + a_2 x^2 + a_3 x^3) = a_1 + 2a_2 x + 3a_3 x^2$$

the corresponding map $\tilde{D}$ is given by the equation

$$\tilde{D} \begin{bmatrix} a_0 \\ a_1 \\ a_2 \\ a_3 \end{bmatrix} = \begin{bmatrix} a_1 \\ 2a_2 \\ 3a_3 \end{bmatrix}$$

The columns of the standard matrix for D are the images of the basis vectors—1, x, x^2, and x^3—of the domain. Because $D(1) = 0$, $D(x) = 1$, $D(x^2) = 2x$, and $D(x^3) = 3x^2$, or, in terms of $\tilde{D}$,

$$\tilde{D} \begin{bmatrix} 1 \\ 0 \\ 0 \\ 0 \end{bmatrix} = \begin{bmatrix} 0 \\ 0 \\ 0 \end{bmatrix}, \quad \tilde{D} \begin{bmatrix} 0 \\ 1 \\ 0 \\ 0 \end{bmatrix} = \begin{bmatrix} 1 \\ 0 \\ 0 \end{bmatrix}, \quad \tilde{D} \begin{bmatrix} 0 \\ 0 \\ 1 \\ 0 \end{bmatrix} = \begin{bmatrix} 0 \\ 2 \\ 0 \end{bmatrix}, \quad \text{and} \quad \tilde{D} \begin{bmatrix} 0 \\ 0 \\ 0 \\ 1 \end{bmatrix} = \begin{bmatrix} 0 \\ 0 \\ 3 \end{bmatrix}$$

the standard matrix for this differentiation map is

$$[D] = [\tilde{D}] = \begin{bmatrix} 0 & 1 & 0 & 0 \\ 0 & 0 & 2 & 0 \\ 0 & 0 & 0 & 3 \end{bmatrix}$$

Here's an illustration of this result. Consider the polynomial $\mathbf{p} = 4 - 5x + x^2 - 2x^3$, whose derivative is $\mathbf{p}' = -5 + 2x - 6x^2$. In terms of the matrix above,

$$D(\mathbf{p}) = \begin{bmatrix} 0 & 1 & 0 & 0 \\ 0 & 0 & 2 & 0 \\ 0 & 0 & 0 & 3 \end{bmatrix} \begin{bmatrix} 4 \\ -5 \\ 1 \\ -2 \end{bmatrix} = \begin{bmatrix} -5 \\ 2 \\ -6 \end{bmatrix}$$

which corresponds to the polynomial $-5 + 2x - 6x^2$, as expected. ∎

Example 16: Let $T : \mathbf{R}^n \rightarrow \mathbf{R}^n$ be a linear operator and assume that $\mathbf{R}^n$ is being considered with a nonstandard basis. While it is still true that the transformation can be written in terms of multiplication by a matrix, *this matrix will not be the standard matrix*. In fact, one of the reasons for choosing a nonstandard basis for $\mathbf{R}^n$ in the first place is to obtain a simple representative matrix, possibly a diagonal one. Let $B = \{\mathbf{b}_1, \mathbf{b}_2, \ldots, \mathbf{b}_n\}$ be a basis for $\mathbf{R}^n$; then, the representative matrix of an operator T has this property: It multiplies the component vector of $\mathbf{x}$ to give the component vector of $T(\mathbf{x})$:

$$[T]_B[\mathbf{x}]_B = [T(\mathbf{x})]_B \qquad (*)$$

The matrix $[T]_B$ is called the matrix of T **relative to the basis** B. The columns of $[T]_B$ are the component vectors of the images of the basis vectors:

$$[T]_B = \begin{bmatrix} | & | & & | \\ [T(\mathbf{b}_1)]_B & [T(\mathbf{b}_2)]_B & \cdots & [T(\mathbf{b}_n)]_B \\ | & | & & | \end{bmatrix}$$

The matrix for T will be the standard matrix only when the basis for $\mathbf{R}^n$ is the standard basis.

In Example 10, $\mathbf{R}^2$ was considered with the basis $B = \{\mathbf{b}_1 = (1, 1)^T, \mathbf{b}_2 = (1, -3)^T\}$. Because T mapped $\mathbf{b}_1$ to itself, $[T(\mathbf{b}_1)]_B = (1, 0)^T$, and, because T mapped $\mathbf{b}_2$ to $5\mathbf{b}_2$, $[T(\mathbf{b}_2)]_B = (0, 5)^T$. Therefore, the matrix for T relative to the nonstandard basis B is the simple, *diagonal* matrix

$$[T]_B = \begin{bmatrix} 1 & 0 \\ 0 & 5 \end{bmatrix}$$

Recall that the vector $\mathbf{v} = (3, 5)^T$ has components $k_1 = 7/2$ and $k_2 = -1/2$ relative to B: that is, $[\mathbf{v}]_B = (7/2, -1/2)^T$. Equation $(*)$ above then gives

$$[T]_B[\mathbf{v}]_B = \begin{bmatrix} 1 & 0 \\ 0 & 5 \end{bmatrix}\begin{bmatrix} \frac{7}{2} \\ -\frac{1}{2} \end{bmatrix} = \begin{bmatrix} \frac{7}{2} \\ -\frac{5}{2} \end{bmatrix} = [T(\mathbf{v})]_B$$

Since $[T(\mathbf{v})]_B = (7/2, -5/2)^T$, the image of $\mathbf{v}$ under T is

$$\begin{aligned} T(\mathbf{v}) &= \tfrac{7}{2}\mathbf{b}_1 - \tfrac{5}{2}\mathbf{b}_2 \\ &= \tfrac{7}{2}\begin{bmatrix} 1 \\ 1 \end{bmatrix} - \tfrac{5}{2}\begin{bmatrix} 1 \\ -3 \end{bmatrix} \\ &= \begin{bmatrix} 1 \\ 11 \end{bmatrix} \end{aligned}$$

which agrees with the result of Example 10. ∎

The Kernel and Range of a Linear Transformation

Let $T: \mathbf{R}^n \to \mathbf{R}^m$ be a linear transformation; then, the set of all vectors $\mathbf{v}$ in $\mathbf{R}^n$ which T sends to the zero vector in $\mathbf{R}^m$ is called the **kernel** of T and denoted ker T:

$$\ker T = \left\{ \mathbf{v} \in \mathbf{R}^n : T(\mathbf{v}) = \mathbf{0} \right\}$$

Since every linear transformation from $\mathbf{R}^n$ to $\mathbf{R}^m$ is given by multiplication by some m by n matrix A—that is, there exists an m by n matrix A such that $T(\mathbf{x}) = A\mathbf{x}$ for every $\mathbf{x}$ in $\mathbf{R}^n$—the condition for $\mathbf{v}$ to be in the kernel of T can be rewritten $A\mathbf{v} = \mathbf{0}$. Those vectors $\mathbf{v}$ such that $A\mathbf{v} = \mathbf{0}$ form the nullspace of A; therefore, *the kernel of T is the same as the nullspace of A*:

$$\ker T = N(A)$$

Thus, for $T: \mathbf{R}^n \to \mathbf{R}^m$, ker T is a subspace of $\mathbf{R}^n$, and its dimension is called the **nullity** of T. This agrees with the definition of the term *nullity* as the dimension of the nullspace of A.

The **range** of a linear transformation $T: \mathbf{R}^n \to \mathbf{R}^m$ is the collection of all images of T:

$$\text{range}(T) = \left\{ \mathbf{w} \in \mathbf{R}^m : \mathbf{w} = T(\mathbf{v}) \text{ for some } \mathbf{v} \text{ in } \mathbf{R}^n \right\}$$

The range of T also has a description in terms of its matrix representative. A vector $\mathbf{w}$ in $\mathbf{R}^m$ is in the range of T precisely when there exists a vector $\mathbf{v}$ such that $T(\mathbf{v}) = \mathbf{w}$. If $T(\mathbf{v})$ is always equal to $A\mathbf{v}$ for some matrix A, then $\mathbf{w}$ is in the range of T if and only if the equation $A\mathbf{v} = \mathbf{w}$ can be solved for $\mathbf{v}$. *But this is precisely the condition for $\mathbf{w}$ to be in the column space of A.* Therefore, if $T: \mathbf{R}^n \to \mathbf{R}^m$, then the range of T, denoted

$R(T)$, is a subspace of $\mathbf{R}^m$ and, for $T(\mathbf{v}) = A\mathbf{v}$, is the same as the column space of A:

$$R(T) = CS(A)$$

Thus, for $T: \mathbf{R}^n \to \mathbf{R}^m$, $R(T)$ is a subspace of $\mathbf{R}^m$, and its dimension is called the **rank** of T. This agrees with the definition of the term *rank* as the dimension of the column space of A.

Example 17: Determine the kernel, nullity, range, and rank of the linear operator $T: \mathbf{R}^2 \to \mathbf{R}^2$ defined by the equation

$$T\begin{bmatrix} x_1 \\ x_2 \end{bmatrix} = \begin{bmatrix} 2x_1 - x_2 \\ -3x_1 + 4x_2 \end{bmatrix}$$

Since

$$\begin{bmatrix} 2x_1 - x_2 \\ -3x_1 + 4x_2 \end{bmatrix} = \begin{bmatrix} 2 & -1 \\ -3 & 4 \end{bmatrix}\begin{bmatrix} x_1 \\ x_2 \end{bmatrix}$$

it is clear that the standard matrix for T is

$$A = \begin{bmatrix} 2 & -1 \\ -3 & 4 \end{bmatrix}$$

Now, because A is a square matrix with nonzero determinant, it is invertible, and there are two consequences:

(1) The equation $A\mathbf{x} = \mathbf{0}$ has only the trivial solution, so the nullspace of A, which is the kernel of T, is just the trivial subspace, $\{\mathbf{0}\}$. Thus, nullity $T = 0$.

(2) The equation $A\mathbf{x} = \mathbf{b}$ has a (unique) solution for every $\mathbf{b}$ in $\mathbf{R}^2$, so the column space of A, which is the range of T, is all of $\mathbf{R}^2$. Thus, rank $T = 2$. ■

The observations made in the solution of the preceding example are completely general:

Theorem K. Let $T : \mathbf{R}^n \to \mathbf{R}^n$ be a linear operator. Then

$$\ker T = \{\mathbf{0}\} \text{ and } R(T) = \mathbf{R}^n$$

if and only if the standard matrix for T is invertible.

Example 18: Determine the kernel, nullity, range, and rank of the linear operator $T : \mathbf{R}^3 \to \mathbf{R}^3$ given by the equation $T(\mathbf{x}) = A\mathbf{x}$, where

$$A = \begin{bmatrix} 1 & 2 & 3 \\ 4 & 5 & 6 \\ 7 & 8 & 9 \end{bmatrix}$$

Since the determinant of A is zero, the preceding theorem guarantees that the kernel of T is nontrivial, and the range of T is not all of $\mathbf{R}^3$. The augmented matrix $[A|\mathbf{b}]$ may be row-reduced as follows:

$$[A|\mathbf{b}] = \begin{bmatrix} 1 & 2 & 3 & b_1 \\ 4 & 5 & 6 & b_2 \\ 7 & 8 & 9 & b_3 \end{bmatrix}$$

$$\xrightarrow[\substack{-4\mathbf{r}_1 \text{ added to } \mathbf{r}_2 \\ -7\mathbf{r}_1 \text{ added to } \mathbf{r}_3}]{} \begin{bmatrix} 1 & 2 & 3 & b_1 \\ 0 & -3 & -6 & -4b_1 + b_2 \\ 0 & -6 & -12 & -7b_1 + b_3 \end{bmatrix}$$

$$\xrightarrow[-2\mathbf{r}_2 \text{ added to } \mathbf{r}_3]{} \begin{bmatrix} 1 & 2 & 3 & b_1 \\ 0 & -3 & -6 & -4b_1 + b_2 \\ 0 & 0 & 0 & b_1 - 2b_2 + b_3 \end{bmatrix}$$

$$\xrightarrow[(-1/3)\mathbf{r}_2]{} \begin{bmatrix} 1 & 2 & 3 & b_1 \\ 0 & 1 & 2 & -\frac{1}{3}(-4b_1 + b_2) \\ 0 & 0 & 0 & b_1 - 2b_2 + b_3 \end{bmatrix} = [A'|\mathbf{b}']$$

The row of zeros implies that $T(\mathbf{x}) = A\mathbf{x} = \mathbf{b}$ has a solution only for those vectors $\mathbf{b} = (b_1, b_2, b_3)^T$ such that $b_1 - 2b_2 + b_3 = 0$; this describes the column space of A, which is the range of T:

$$R(T) = \left\{ \mathbf{b} = (b_1, b_2, b_3)^T \in \mathbf{R}^3 : b_1 - 2b_2 + b_3 = 0 \right\}$$

and it follows from the two nonzero rows in the echelon form of A obtained above that rank A = rank $T = 2$.

It also follows from the row reduction above that the set of solutions of $A\mathbf{x} = \mathbf{0}$ is identical to the set of solutions of $A'\mathbf{x} = \mathbf{0}$, where

$$A' = \begin{bmatrix} 1 & 2 & 3 \\ 0 & 1 & 2 \\ 0 & 0 & 0 \end{bmatrix}$$

Let $\mathbf{x} = (x_1, x_2, x_3)^T$. The two nonzero rows in A' imply that 3 − 2 = 1 of the variables is free; let x_3 be the free variable. Then back-substitution into the second row yields $x_2 = -2x_3$, and back-substitution into the first row gives $x_1 = x_3$. The vectors $\mathbf{x}$ that satisfy $A\mathbf{x} = \mathbf{0}$ are those of the form $\mathbf{x} = (x_3, -2x_3, x_3) = x_3(1\ -2,\ 1)$. Therefore, the nullspace of A—the kernel of T—is

$$\ker T = N(A) = \left\{ \mathbf{x} \in \mathbf{R}^3 : \mathbf{x} = t(1,\ -2,\ 1)^T \text{ for any } t \text{ in } \mathbf{R} \right\}$$

This is a 1-dimensional subspace of $\mathbf{R}^3$, so nullity $T = 1$. ∎

Note that the rank plus nullity theorem continues to hold in this setting of linear maps $T : \mathbf{R}^n \to \mathbf{R}^m$, that is,

$$\boxed{\text{rank}(T) + \text{nullity}(T) = n = \dim (\text{domain of } T)}$$

Example 19: Determine the kernel, nullity, range, and rank of the linear map $T : \mathbf{R}^3 \to \mathbf{R}^2$ defined by the equation $T(\mathbf{x}) = A\mathbf{x}$, where

$$A = \begin{bmatrix} 1 & -2 & 1 \\ 2 & -3 & -4 \end{bmatrix}$$

Theorem K does not apply here, since T is not a linear operator; the matrix for T is not square. One elementary row operation reduces A to echelon form:

$$A = \begin{bmatrix} 1 & -2 & 1 \\ 2 & -3 & -4 \end{bmatrix} \xrightarrow{-2r_1 \text{ added to } r_2} \begin{bmatrix} 1 & -2 & 1 \\ 0 & 1 & -6 \end{bmatrix} = A'$$

To find the kernel of T, the equation $A'\mathbf{x} = \mathbf{0}$ must be solved. Let $\mathbf{x} = (x_1, x_2, x_3)^{\text{T}}$. The two nonzero rows in A' imply $3 - 2 = 1$ of the variables is free; let x_3 be the free variable. The second row implies $x_2 = 6x_3$, and back-substitution into the first row gives $x_1 = 11x_3$. The vectors $\mathbf{x}$ that satisfy $A\mathbf{x} = \mathbf{0}$ are those of the form $\mathbf{x} = (11x_3, 6x_3, x_3)^{\text{T}} = x_3(11, 6, 1)^{\text{T}}$. Therefore, the nullspace of A—the kernel of T—is

$$\ker T = N(A) = \left\{ \mathbf{x} \in \mathbf{R}^3 : \mathbf{x} = t(11, 6, 1)^{\text{T}} \text{ for any } t \text{ in } \mathbf{R} \right\}$$

This is a 1-dimensional subspace of $\mathbf{R}^3$, so nullity $T = 1$.

Now, by the rank plus nullity theorem,
$$\text{rank}(T) + \text{nullity}(T) = n = \dim(\text{domain } T)$$
$$\text{rank}(T) + 1 = 3$$
$$\text{rank}(T) = 2$$

Because T maps vectors into $\mathbf{R}^2$, the range of T is a subspace of dimension 2 (since rank $T = 2$) of $\mathbf{R}^2$. Since the only 2-dimensional subspace of $\mathbf{R}^2$ is $\mathbf{R}^2$ itself, $R(T) = \mathbf{R}^2$. ∎

Example 19 illustrates an important fact about some linear maps that are not linear operators:

Theorem L. Let $T : \mathbf{R}^n \to \mathbf{R}^m$ be a linear map with $m < n$. Since the standard matrix of T—which is m by n—has fewer rows than columns, the kernel of T cannot be the trivial subspace of $\mathbf{R}^n$. In fact, nullity$(T) \geq n - m > 0$.

Example 20: Determine the kernel, nullity, range, and rank of the linear map $T: \mathbf{R}^2 \rightarrow \mathbf{R}^3$ defined by the equation $T(\mathbf{x}) = A\mathbf{x}$, where

$$A = \begin{bmatrix} 1 & -2 \\ 2 & -5 \\ -1 & 0 \end{bmatrix}$$

Gaussian elimination transforms A into echelon form:

$$A = \begin{bmatrix} 1 & -2 \\ 2 & -5 \\ -1 & 0 \end{bmatrix} \xrightarrow[\mathbf{r}_1 \text{ added to } \mathbf{r}_3]{-2\mathbf{r}_1 \text{ added to } \mathbf{r}_2} \begin{bmatrix} 1 & -2 \\ 0 & -1 \\ 0 & -2 \end{bmatrix}$$

$$\xrightarrow{-2\mathbf{r}_2 \text{ added to } \mathbf{r}_3} \begin{bmatrix} 1 & -2 \\ 0 & -1 \\ 0 & 0 \end{bmatrix} = A'$$

Since there are just two columns, the fact that there are two nonzero rows in A' implies that there are $2 - 2 = 0$ free variables in the solution of $A\mathbf{x} = \mathbf{0}$. Since a homogeneous system has either infinitely many solutions or just the trivial solution, the absence of any free variables means that the only solution is $\mathbf{x} = \mathbf{0}$. Thus, the kernel of T is trivial: ker $T = \{\mathbf{0}\} \subset \mathbf{R}^2$ and nullity $T = 0$.

Now, the rank plus nullity theorem,

$$\text{rank}(T) + \text{nullity}(T) = n = \dim(\text{domain } T)$$
$$\text{rank}(T) + 0 = 2$$
$$\text{rank}(T) = 2$$

Since T maps vectors into $\mathbf{R}^3$, the range of T is a subspace of $\mathbf{R}^3$ of dimension 2. It follows directly from the definition of T that

$$R(T) = \left\{ T(\mathbf{x}) : \mathbf{x} \in \mathbf{R}^2 \right\}$$

$$= \left\{ A\mathbf{x} : \mathbf{x} \in \mathbf{R}^2 \right\}$$

$$= \left\{ \begin{bmatrix} 1 & -2 \\ 2 & -5 \\ -1 & 0 \end{bmatrix} \begin{bmatrix} x_1 \\ x_2 \end{bmatrix} : x_1, x_2 \in \mathbf{R} \right\}$$

$$= \left\{ x_1 \begin{bmatrix} 1 \\ 2 \\ -1 \end{bmatrix} + x_2 \begin{bmatrix} -2 \\ -5 \\ 0 \end{bmatrix} : x_1, x_2 \in \mathbf{R} \right\}$$

Since every 2-dimensional subspace of $\mathbf{R}^3$ is a plane through the origin, the range of T can be expressed as such a plane. Since $R(T)$ contains $\mathbf{w}_1 = (1, 2, -1)^T = \mathbf{i} + 2\mathbf{j} - \mathbf{k}$ and $\mathbf{w}_2 = (-2, -5, 0)^T = -2\mathbf{i} - 5\mathbf{j} + 0\mathbf{k}$, a normal vector to this plane is

$$\mathbf{n} = \mathbf{w}_1 \times \mathbf{w}_2 = \begin{vmatrix} \mathbf{i} & \mathbf{j} & \mathbf{k} \\ 1 & 2 & -1 \\ -2 & -5 & 0 \end{vmatrix}$$

$$= \mathbf{i} \begin{vmatrix} 2 & -1 \\ -5 & 0 \end{vmatrix} - \mathbf{j} \begin{vmatrix} 1 & -1 \\ -2 & 0 \end{vmatrix} + \mathbf{k} \begin{vmatrix} 1 & 2 \\ -2 & -5 \end{vmatrix}$$

$$= -5\mathbf{i} + 2\mathbf{j} - \mathbf{k}$$

The standard equation for the plane is therefore $-5x + 2y - z = d$ for some constant d. Since this plane must contain the origin (it's a subspace), d must be 0. Thus, the range of T can also be written in the form

$$R(T) = \left\{ (x,\, y,\, z)^{\mathrm{T}} :\ 5x - 2y + z = 0 \right\} \quad \blacksquare$$

Injectivity and surjectivity. A linear transformation $T: \mathbf{R}^n \to \mathbf{R}^m$ is said to be **one to one** (or **injective**) if no two vectors in $\mathbf{R}^n$ are mapped to the same vector in $\mathbf{R}^m$. That is, T is one to one if and only if

$$T(\mathbf{v}_1) = T(\mathbf{v}_2) \quad \Rightarrow \quad \mathbf{v}_1 = \mathbf{v}_2 \quad (*)$$

A map $T : \mathbf{R}^n \to \mathbf{R}^m$ is said to be **onto** (or **surjective**) if the range of T is all of $\mathbf{R}^m$.

Theorem M. A linear map $T : \mathbf{R}^n \to \mathbf{R}^m$ is one to one if and only if ker $T = \{\mathbf{0}\}$.

Proof. ($\Rightarrow$) First, assume that T is one to one. If $\mathbf{v}$ is in ker T, then $T(\mathbf{v}) = \mathbf{0}$. Since $T(\mathbf{0}) = \mathbf{0}$ also, $T(\mathbf{v}) = T(\mathbf{0}) = \mathbf{0}$, which, applying (*), implies $\mathbf{v} = \mathbf{0}$. Therefore, ker T contains only the zero vector.

($\Leftarrow$) Now, assume that T is not one to one, that is, assume there exist *distinct* vectors $\mathbf{v}_1$ and $\mathbf{v}_2$ in $\mathbf{R}^n$ such that $T(\mathbf{v}_1) = T(\mathbf{v}_2)$. Then, $T(\mathbf{v}_1) - T(\mathbf{v}_2) = \mathbf{0}$, which, by the linearity of T, implies $T(\mathbf{v}_1 - \mathbf{v}_2) = \mathbf{0}$. Since $\mathbf{v}_1 \neq \mathbf{v}_2$, T maps the *nonzero* vector $\mathbf{v}_1 - \mathbf{v}_2$ to $\mathbf{0}$; therefore, ker $T \neq \{\mathbf{0}\}$. $\quad \blacksquare$

Example 21: Consider the linear map $T : \mathbf{R}^3 \to \mathbf{R}^2$ defined by the equation $T(\mathbf{x}) = A\mathbf{x}$, where

$$A = \begin{bmatrix} 1 & -2 & 1 \\ 2 & -3 & -4 \end{bmatrix}$$

(This is the map given in Example 19.) Is T one to one? Is it onto?

In Example 19, it was shown that

$$\ker T = N(A) = \left\{ \mathbf{x} \in \mathbf{R}^3 : \mathbf{x} = t(11,\ 6,\ 1)^T \text{ for any } t \text{ in } \mathbf{R} \right\}$$

which is a line through the origin. Since $\ker T$ contains vectors other than $\mathbf{0}$, T is *not* one to one. However, since rank T = rank $A = 2$, the range is all of $\mathbf{R}^2$, so T is onto. ∎

Assume that $T : \mathbf{R}^n \to \mathbf{R}^n$ is a linear operator which is one to one. Then $\ker T = \{\mathbf{0}\}$, so the nullity of T is 0. The rank plus nullity theorem then guarantees that rank T is $n - 0 = n$, which means the range of T is an n-dimensional subspace of $\mathbf{R}^n$. But the only n-dimensional subspace of $\mathbf{R}^n$ is $\mathbf{R}^n$ itself, which means that T is onto. This argument justifies the following result:

Theorem N. A linear operator $T : \mathbf{R}^n \to \mathbf{R}^n$ is one to one if and only if it is onto, and this case arises precisely when the standard matrix for T is invertible, that is, when det $[T] \neq 0$.

Example 22: Since det $A \neq 0$, Theorem N guarantees that the linear operator $T : \mathbf{R}^2 \to \mathbf{R}^2$ defined by the equation $T(\mathbf{x}) = A\mathbf{x}$, where

$$A = \begin{bmatrix} 2 & -1 \\ -3 & 4 \end{bmatrix}$$

is both one to one and onto; therefore, $\ker T = \{\mathbf{0}\}$ and $R(T) = \mathbf{R}^2$.

The linear operator $T: \mathbf{R}^3 \to \mathbf{R}^3$ defined by the equation $T(\mathbf{x}) = A\mathbf{x}$, where

$$A = \begin{bmatrix} 1 & 2 & 3 \\ 4 & 5 & 6 \\ 7 & 8 & 9 \end{bmatrix}$$

is neither one to one nor onto, since det A does equal zero. The range (which is *not* all of $\mathbf{R}^3$) and the nontrivial kernel of this operator were explicitly determined in Example 18. ∎

Example 23: Define an operator $P: \mathbf{R}^3 \to \mathbf{R}^3$ by the formula $P(x, y, z)^{\mathrm{T}} = (x, y, 0)^{\mathrm{T}}$. Is P one to one? Is it onto?

The effect of this map is to project every point in $\mathbf{R}^3$ onto the x-y plane; see Figure 54. Intuitively, then, P cannot be one to one: To illustrate, both $(1, 2, 3)^{\mathrm{T}}$ and $(1, 2, 4)^{\mathrm{T}}$ get mapped to $(1, 2, 0)^{\mathrm{T}}$. Furthermore, it cannot be onto $\mathbf{R}^3$, since the image of every point lies in the x-y plane only; this is the range of P. Note that this is consistent with Theorem N, since an operator is one to one if and only if it is onto, and the standard matrix for P,

$$[P] = \begin{bmatrix} 1 & 0 & 0 \\ 0 & 1 & 0 \\ 0 & 0 & 0 \end{bmatrix}$$

has determinant 0.

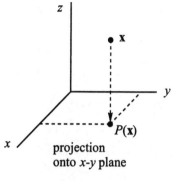

projection
onto x-y plane

■ Figure 54 ■

Example 24: Let $T:\mathbf{R}^n \to \mathbf{R}^n$ be a linear operator which is both one to one and onto. Then T has an **inverse**, $T^{-1}:\mathbf{R}^n \to \mathbf{R}^n$, which is defined as follows: If $T(\mathbf{x}) = \mathbf{y}$, then $T^{-1}(\mathbf{y}) = \mathbf{x}$. Since T is both one to one and onto, its standard matrix, $[T]$, is invertible; *its inverse, $[T]^{-1}$, is the standard matrix for T^{-1}*:

$$\boxed{[T^{-1}] = [T]^{-1}}$$

Consider the linear operator $T:\mathbf{R}^2 \to \mathbf{R}^2$ defined by the equation

$$T(\mathbf{x}) = \begin{bmatrix} 2 & -1 \\ -3 & 4 \end{bmatrix} \mathbf{x}$$

This operator appeared in Example 11, where it was calculated that $T(3, 5)^\mathrm{T} = (1, 11)^\mathrm{T}$. Obtain a formula for the inverse of T and verify that $T^{-1}(1, 11)^\mathrm{T} = (3, 5)^\mathrm{T}$.

Since the standard matrix for T has a nonzero determinant, it is invertible, and its inverse,

$$[T]^{-1} = \begin{bmatrix} 2 & -1 \\ -3 & 4 \end{bmatrix}^{-1} = \tfrac{1}{5} \begin{bmatrix} 4 & 1 \\ 3 & 2 \end{bmatrix}$$

is the standard matrix for the inverse operator T^{-1}. Therefore,

$$T^{-1}(\mathbf{x}) = \tfrac{1}{5} \begin{bmatrix} 4 & 1 \\ 3 & 2 \end{bmatrix} \mathbf{x}$$

and

$$T^{-1}\begin{bmatrix} 1 \\ 11 \end{bmatrix} = \tfrac{1}{5} \begin{bmatrix} 4 & 1 \\ 3 & 2 \end{bmatrix}\begin{bmatrix} 1 \\ 11 \end{bmatrix} = \tfrac{1}{5}\begin{bmatrix} 15 \\ 25 \end{bmatrix} = \begin{bmatrix} 3 \\ 5 \end{bmatrix}$$

as expected. ■

Example 25: Find the inverse of the rotation operator T_θ given in Example 14.

Since T_θ rotates every vector through the angle θ, the inverse operator should rotate it back; that is, $(T_\theta)^{-1}$ should rotate every vector through an angle of $-\theta$. Since the standard matrix for T_θ is A_θ, the standard matrix for $(T_\theta)^{-1}$ is $(A_\theta)^{-1}$, which by the argument just given, is equal to $A_{-\theta}$:

$$(A_\theta)^{-1} = A_{-\theta} = \begin{bmatrix} \cos(-\theta) & -\sin(-\theta) \\ \sin(-\theta) & \cos(-\theta) \end{bmatrix} = \begin{bmatrix} \cos\theta & \sin\theta \\ -\sin\theta & \cos\theta \end{bmatrix}$$

Note that this intuitive, geometric argument gives the same result as would be obtained by formally taking the inverse of the 2 by 2 matrix A_θ. ■

Composition of Linear Transformations

Let V, W, and Z be vector spaces and let $T_1 \colon V \to W$ and $T_2 \colon W \to Z$ be linear transformations. The **composition** of T_1 and T_2, denoted $T_2 \circ T_1$, is defined to be the linear transformation from V to Z given by the equation

$$T_2 \circ T_1(\mathbf{v}) = T_2(T_1(\mathbf{v}))$$

The notation $T_2 \circ T_1$ is meant to be read from right to left; that is, first apply T_1 and then T_2. The composition $T_2 \circ T_1$ maps V all the way to Z; see Figure 55.

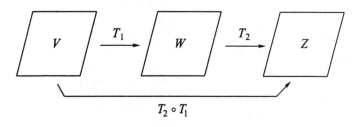

$$T_2 \circ T_1$$

■ Figure 55 ■

Example 26: Consider the linear transformations $T_1 \colon \mathbf{R}^3 \to \mathbf{R}^2$ and $T_2 \colon \mathbf{R}^2 \to \mathbf{R}^4$ defined by the equations

$$T_1\begin{bmatrix} v_1 \\ v_2 \\ v_3 \end{bmatrix} = \begin{bmatrix} v_1 - 2v_2 \\ -3v_1 + v_2 - v_3 \end{bmatrix} \quad \text{and} \quad T_2\begin{bmatrix} w_1 \\ w_2 \end{bmatrix} = \begin{bmatrix} 4w_2 \\ 2w_1 - w_2 \\ -w_1 + w_2 \\ 5w_1 \end{bmatrix}$$

Find a formula for the composition $T_2 \circ T_1 \colon \mathbf{R}^3 \to \mathbf{R}^4$.

From the definitions of T_1 and T_2,

$$T_2(T_1(\mathbf{v})) = T_2 \begin{bmatrix} v_1 - 2v_2 \\ -3v_1 + v_2 - v_3 \end{bmatrix}$$

$$= \begin{bmatrix} 4(-3v_1 + v_2 - v_3) \\ 2(v_1 - 2v_2) - (-3v_1 + v_2 - v_3) \\ -(v_1 - 2v_2) + (-3v_1 + v_2 - v_3) \\ 5(v_1 - 2v_2) \end{bmatrix}$$

$$= \begin{bmatrix} -12v_1 + 4v_2 - 4v_3 \\ 5v_1 - 5v_2 + v_3 \\ -4v_1 + 3v_2 - v_3 \\ 5v_1 - 10v_2 \end{bmatrix}$$

Therefore,

$$(T_2 \circ T_1) \begin{bmatrix} v_1 \\ v_2 \\ v_3 \end{bmatrix} = \begin{bmatrix} -12v_1 + 4v_2 - 4v_3 \\ 5v_1 - 5v_2 + v_3 \\ -4v_1 + 3v_2 - v_3 \\ 5v_1 - 10v_2 \end{bmatrix} \quad \blacksquare$$

Example 27: For the linear transformations T_1 and T_2 in Example 26 above, find the standard matrix representatives for T_1, T_2, and $T_2 \circ T_1$, then show that $[T_2 \circ T_1] = [T_2][T_1]$.

The standard matrix for $T_1 \colon \mathbf{R}^3 \to \mathbf{R}^2$ is the 2×3 matrix whose columns are the images of the basis vectors of the domain space, $\mathbf{R}^3$. Since

$$T_1 \begin{bmatrix} 1 \\ 0 \\ 0 \end{bmatrix} = \begin{bmatrix} 1 \\ -3 \end{bmatrix}, \quad T_1 \begin{bmatrix} 0 \\ 1 \\ 0 \end{bmatrix} = \begin{bmatrix} -2 \\ 1 \end{bmatrix}, \quad \text{and} \quad T_1 \begin{bmatrix} 0 \\ 0 \\ 1 \end{bmatrix} = \begin{bmatrix} 0 \\ -1 \end{bmatrix}$$

the standard matrix for T_1 is

$$[T_1] = \begin{bmatrix} 1 & -2 & 0 \\ -3 & 1 & -1 \end{bmatrix}$$

The standard matrix for $T_2 : \mathbf{R}^2 \to \mathbf{R}^4$ is the 4×2 matrix whose columns are the images of the basis vectors of $\mathbf{R}^2$. Since

$$T_2 \begin{bmatrix} 1 \\ 0 \end{bmatrix} = \begin{bmatrix} 0 \\ 2 \\ -1 \\ 5 \end{bmatrix} \quad \text{and} \quad T_2 \begin{bmatrix} 0 \\ 1 \end{bmatrix} = \begin{bmatrix} 4 \\ -1 \\ 1 \\ 0 \end{bmatrix}$$

the standard matrix for T_2 is

$$[T_2] = \begin{bmatrix} 0 & 4 \\ 2 & -1 \\ -1 & 1 \\ 5 & 0 \end{bmatrix}$$

Finally, the standard matrix for $T_2 \circ T_1 : \mathbf{R}^3 \to \mathbf{R}^4$ is the 4×3 matrix whose columns are the images of the standard basis vectors of $\mathbf{R}^3$. From the result of Example 26,

$$(T_2 \circ T_1) \begin{bmatrix} 1 \\ 0 \\ 0 \end{bmatrix} = \begin{bmatrix} -12 \\ 5 \\ -4 \\ 5 \end{bmatrix}, \quad (T_2 \circ T_1) \begin{bmatrix} 0 \\ 1 \\ 0 \end{bmatrix} = \begin{bmatrix} 4 \\ -5 \\ 3 \\ -10 \end{bmatrix},$$

$$\text{and} \quad (T_2 \circ T_1) \begin{bmatrix} 0 \\ 0 \\ 1 \end{bmatrix} = \begin{bmatrix} -4 \\ 1 \\ -1 \\ 0 \end{bmatrix}$$

the standard matrix for $T_2 \circ T_1$ is

$$[T_2 \circ T_1] = \begin{bmatrix} -12 & 4 & -4 \\ 5 & -5 & 1 \\ -4 & 3 & -1 \\ 5 & -10 & 0 \end{bmatrix}$$

Now, verify the following matrix multiplication:

$$[T_2][T_1] = \begin{bmatrix} 0 & 4 \\ 2 & -1 \\ -1 & 1 \\ 5 & 0 \end{bmatrix} \begin{bmatrix} 1 & -2 & 0 \\ -3 & 1 & -1 \end{bmatrix} = \begin{bmatrix} -12 & 4 & -4 \\ 5 & -5 & 1 \\ -4 & 3 & -1 \\ 5 & -10 & 0 \end{bmatrix}$$

This calculation shows that the standard matrix for $T_2 \circ T_1$ is the product of the standard matrices for T_2 and T_1:

$$\boxed{[T_2 \circ T_1] = [T_2][T_1]}$$

It can be shown that this equation holds true for any linear transformations T_1 and T_2 for which $T_2 \circ T_1$ is defined, and is, in fact, the motivation behind the rather involved definition of matrix multiplication. ∎

Example 28: Let T_1 and T_2 be the linear operators on $\mathbf{R}^2$ defined by the equations

$$T_1\begin{bmatrix} x_1 \\ x_2 \end{bmatrix} = \begin{bmatrix} 2x_1 - 3x_2 \\ -x_1 + 4x_2 \end{bmatrix} \quad \text{and} \quad T_2\begin{bmatrix} x_1 \\ x_2 \end{bmatrix} = \begin{bmatrix} -4x_2 \\ x_1 + x_2 \end{bmatrix}$$

Compute the compositions $T_1 \circ T_2$ and $T_2 \circ T_1$. Does $T_1 \circ T_2 = T_2 \circ T_1$?

The composition $T_1 \circ T_2$ is given by the formula

$$\begin{aligned}
T_1 \circ T_2(\mathbf{x}) &= T_1(T_2(\mathbf{x})) \\
&= T_1\begin{bmatrix} -4x_2 \\ x_1 + x_2 \end{bmatrix} \\
&= \begin{bmatrix} 2(-4x_2) - 3(x_1 + x_2) \\ -(-4x_2) + 4(x_1 + x_2) \end{bmatrix} \\
&= \begin{bmatrix} -3x_1 - 11x_2 \\ 4x_1 + 8x_2 \end{bmatrix}
\end{aligned}$$

while the composition $T_2 \circ T_1$ is given by the formula

$$\begin{aligned}
T_2 \circ T_1(\mathbf{x}) &= T_2(T_1(\mathbf{x})) \\
&= T_2\begin{bmatrix} 2x_1 - 3x_2 \\ -x_1 + 4x_2 \end{bmatrix} \\
&= \begin{bmatrix} -4(-x_1 + 4x_2) \\ (2x_1 - 3x_2) + (-x_1 + 4x_2) \end{bmatrix} \\
&= \begin{bmatrix} 4x_1 - 16x_2 \\ x_1 + x_2 \end{bmatrix}
\end{aligned}$$

Another method to determine the compositions $T_1 \circ T_2$ and $T_2 \circ T_1$ is to compute the matrix products $[T_1][T_2]$ and $[T_2][T_1]$. Since

$$T_1 \begin{bmatrix} x_1 \\ x_2 \end{bmatrix} = \begin{bmatrix} 2x_1 - 3x_2 \\ -x_1 + 4x_2 \end{bmatrix} \;\Rightarrow\; [T_1] = \begin{bmatrix} 2 & -3 \\ -1 & 4 \end{bmatrix}$$

and

$$T_2 \begin{bmatrix} x_1 \\ x_2 \end{bmatrix} = \begin{bmatrix} -4x_2 \\ x_1 + x_2 \end{bmatrix} \;\Rightarrow\; [T_2] = \begin{bmatrix} 0 & -4 \\ 1 & 1 \end{bmatrix}$$

the matrix products are

$$[T_1][T_2] = \begin{bmatrix} 2 & -3 \\ -1 & 4 \end{bmatrix} \begin{bmatrix} 0 & -4 \\ 1 & 1 \end{bmatrix} = \begin{bmatrix} -3 & -11 \\ 4 & 8 \end{bmatrix}$$

and

$$[T_2][T_1] = \begin{bmatrix} 0 & -4 \\ 1 & 1 \end{bmatrix} \begin{bmatrix} 2 & -3 \\ -1 & 4 \end{bmatrix} = \begin{bmatrix} 4 & -16 \\ 1 & 1 \end{bmatrix}$$

But

$$[T_1][T_2] = \begin{bmatrix} -3 & -11 \\ 4 & 8 \end{bmatrix} \;\Rightarrow\; T_1 \circ T_2(\mathbf{x}) = \begin{bmatrix} -3x_1 - 11x_2 \\ 4x_1 + 8x_2 \end{bmatrix}$$

and

$$[T_2][T_1] = \begin{bmatrix} 4 & -16 \\ 1 & 1 \end{bmatrix} \;\Rightarrow\; T_2 \circ T_1(\mathbf{x}) = \begin{bmatrix} 4x_1 - 16x_2 \\ x_1 + x_2 \end{bmatrix}$$

as above. Clearly, $T_1 \circ T_2 \neq T_2 \circ T_1$. The noncommutativity of linear-map composition is reflected in the noncommutativity of matrix multiplication: in general, $T_1 \circ T_2 \neq T_2 \circ T_1$, since—in general—$[T_1][T_2] \neq [T_2][T_1]$. ■

Example 29: If a linear operator T is composed with itself, the resulting operator is written T^2, rather than $T \circ T$. If T_θ is the rotation operator on $\mathbf{R}^2$ defined in Example 14, find a formula for T_θ^2.

Since T_θ rotates a vector through the angle θ, applying the operator again should rotate the vector through another angle of θ; that is, the effect of T_θ^2 is to rotate every vector through an angle of 2θ. Since the standard matrix for T_θ is A_θ, the standard matrix for T_θ^2 is $A_{2\theta}$:

$$[T_\theta^2] = A_{2\theta} = \begin{bmatrix} \cos 2\theta & -\sin 2\theta \\ \sin 2\theta & \cos 2\theta \end{bmatrix}$$

Now, of course, the standard matrix for T^2 is the square of the standard matrix for T:

$$[T^2] = [T \circ T] = [T][T] = [T]^2$$

So, this same result could have been obtained by squaring the matrix A_θ:

$$\begin{aligned} A_\theta^2 &= \begin{bmatrix} \cos\theta & -\sin\theta \\ \sin\theta & \cos\theta \end{bmatrix}\begin{bmatrix} \cos\theta & -\sin\theta \\ \sin\theta & \cos\theta \end{bmatrix} \\ &= \begin{bmatrix} \cos^2\theta - \sin^2\theta & -2\sin\theta\cos\theta \\ 2\sin\theta\cos\theta & \cos^2\theta - \sin^2\theta \end{bmatrix} \\ &= \begin{bmatrix} \cos 2\theta & -\sin 2\theta \\ \sin 2\theta & \cos 2\theta \end{bmatrix} \end{aligned}$$

where the last equation is a consequence of the trigonometric identities $\cos 2\theta = \cos^2\theta - \sin^2\theta$ and $\sin 2\theta = 2\sin\theta\cos\theta$. Alternatively, you may look at this as a *proof* of these identities. ∎

$\mathbf{A}$lthough the process of applying a linear operator T to a vector gives a vector in the same space as the original, the resulting vector usually points in a completely different direction from the original, that is, $T(\mathbf{x})$ is neither parallel nor antiparallel to $\mathbf{x}$. However, it can happen that $T(\mathbf{x})$ *is* a scalar multiple of $\mathbf{x}$—even when $\mathbf{x} \neq \mathbf{0}$—and this phenomenon is so important that it deserves to be explored.

Definition and Illustration of an Eigenvalue and an Eigenvector

If $T: \mathbf{R}^n \rightarrow \mathbf{R}^n$ is a linear operator, then T must be given by $T(\mathbf{x}) = A\mathbf{x}$ for some $n \times n$ matrix A. If $\mathbf{x} \neq \mathbf{0}$ and $T(\mathbf{x}) = A\mathbf{x}$ is a scalar multiple of $\mathbf{x}$, that is, if

$$A\mathbf{x} = \lambda\mathbf{x}$$

for some scalar λ, then λ is said to be an **eigenvalue** of T (or, equivalently, of A). Any *nonzero* vector $\mathbf{x}$ which satisfies this equation is said to be an **eigenvector** of T (or of A) corresponding to λ. To illustrate these definitions, consider the linear operator $T: \mathbf{R}^2 \rightarrow \mathbf{R}^2$ defined by the equation

$$T(\mathbf{x}) = \begin{bmatrix} 1 & -2 \\ 3 & -4 \end{bmatrix} \mathbf{x}$$

That is, T is given by left multiplication by the matrix

$$A = \begin{bmatrix} 1 & -2 \\ 3 & -4 \end{bmatrix}$$

Consider, for example, the image of the vector $\mathbf{x} = (1, 3)^{\mathrm{T}}$ under the action of T:

$$T\begin{bmatrix}1\\3\end{bmatrix} = \begin{bmatrix}1 & -2\\3 & -4\end{bmatrix}\begin{bmatrix}1\\3\end{bmatrix} = \begin{bmatrix}-5\\-9\end{bmatrix}$$

Clearly, $T(\mathbf{x})$ is not a scalar multiple of $\mathbf{x}$, and this is what typically occurs.

However, now consider the image of the vector $\mathbf{x} = (2, 3)^{\mathrm{T}}$ under the action of T:

$$T\begin{bmatrix}2\\3\end{bmatrix} = \begin{bmatrix}1 & -2\\3 & -4\end{bmatrix}\begin{bmatrix}2\\3\end{bmatrix} = \begin{bmatrix}-4\\-6\end{bmatrix}$$

Here, $T(\mathbf{x})$ *is* a scalar multiple of $\mathbf{x}$, since $T(\mathbf{x}) = (-4, -6)^{\mathrm{T}} = -2(2, 3)^{\mathrm{T}} = -2\mathbf{x}$. Therefore, -2 is an eigenvalue of T, and $(2, 3)^{\mathrm{T}}$ is an eigenvector corresponding to this eigenvalue. The question now is, how do you determine the eigenvalues and associated eigenvectors of a linear operator?

Determining the Eigenvalues of a Matrix

Since every linear operator is given by left multiplication by some square matrix, finding the eigenvalues and eigenvectors of a linear operator is equivalent to finding the eigenvalues and eigenvectors of the associated square matrix; this is the terminology that will be followed. Furthermore, since eigenvalues and eigenvectors make sense only for square matrices, throughout this section all matrices are assumed to be square.

Given a square matrix A, the condition that characterizes an eigenvalue, λ, is the existence of a *nonzero* vector $\mathbf{x}$ such that $A\mathbf{x} = \lambda\mathbf{x}$; this equation can be rewritten as follows:

$$A\mathbf{x} = \lambda\mathbf{x}$$
$$A\mathbf{x} - \lambda\mathbf{x} = \mathbf{0}$$
$$A\mathbf{x} - \lambda I\mathbf{x} = \mathbf{0}$$
$$(A - \lambda I)\mathbf{x} = \mathbf{0}$$

This final form of the equation makes it clear that $\mathbf{x}$ is the solution of a square, homogeneous system. If *nonzero* solutions are desired, then the determinant of the coefficient matrix— which in this case is $A - \lambda I$—must be zero; if not, then the system possesses only the trivial solution $\mathbf{x} = \mathbf{0}$. Since eigenvectors are, by definition, nonzero, in order for $\mathbf{x}$ to be an eigenvector of a matrix A, λ must be chosen so that

$$\det(A - \lambda I) = 0$$

When the determinant of $A - \lambda I$ is written out, the resulting expression is a monic polynomial in λ. [A *monic* polynomial is one in which the coefficient of the leading (the highest-degree) term is 1.] It is called the **characteristic polynomial** of A and will be of degree n if A is $n \times n$. The zeros of the characteristic polynomial of A—that is, the solutions of the **characteristic equation**, $\det(A - \lambda I) = 0$—are the eigenvalues of A.

Example 1: Determine the eigenvalues of the matrix

$$A = \begin{bmatrix} 1 & -2 \\ 3 & -4 \end{bmatrix}$$

First, form the matrix $A - \lambda I$:

$$A - \lambda I = \begin{bmatrix} 1 & -2 \\ 3 & -4 \end{bmatrix} - \begin{bmatrix} \lambda & \\ & \lambda \end{bmatrix} = \begin{bmatrix} 1-\lambda & -2 \\ 3 & -4-\lambda \end{bmatrix}$$

a result which follows by simply subtracting λ from each of the entries on the main diagonal. Now, take the determinant of $A - \lambda I$:

$$\det(A - \lambda I) = \det \begin{bmatrix} 1-\lambda & -2 \\ 3 & -4-\lambda \end{bmatrix} = (1-\lambda)(-4-\lambda) - (3)(-2)$$

$$= \lambda^2 + 3\lambda + 2$$

This is the characteristic polynomial of A, and the solutions of the characteristic equation, $\det(A - \lambda I) = 0$, are the eigenvalues of A:

$$\det(A - \lambda I) = 0$$
$$\lambda^2 + 3\lambda + 2 = 0$$
$$(\lambda + 1)(\lambda + 2) = 0$$
$$\lambda = -1, \; -2 \quad \blacksquare$$

In some texts, the characteristic polynomial of A is written $\det(\lambda I - A)$, rather than $\det(A - \lambda I)$. For matrices of even dimension, these polynomials are precisely the same, while for square matrices of odd dimension, these polynomials are additive inverses. The distinction is merely cosmetic, because the solutions of $\det(\lambda I - A) = 0$ are precisely the same as the solutions of $\det(A - \lambda I) = 0$. Therefore, whether you write the characteristic polynomial of A as $\det(\lambda I - A)$ or as $\det(A - \lambda I)$ will have no effect on the determination of the eigenvalues or their corresponding eigenvectors.

Example 2: Find the eigenvalues of the 3 by 3 checkerboard matrix

$$C = \begin{bmatrix} 1 & -1 & 1 \\ -1 & 1 & -1 \\ 1 & -1 & 1 \end{bmatrix}$$

The determinant

$$\det(C - \lambda I) = \det \begin{bmatrix} 1-\lambda & -1 & 1 \\ -1 & 1-\lambda & -1 \\ 1 & -1 & 1-\lambda \end{bmatrix}$$

is evaluated by first adding the second row to the third and then performing a Laplace expansion by the first column:

$$\begin{vmatrix} 1-\lambda & -1 & 1 \\ -1 & 1-\lambda & -1 \\ 1 & -1 & 1-\lambda \end{vmatrix} = \begin{vmatrix} 1-\lambda & -1 & 1 \\ -1 & 1-\lambda & -1 \\ 0 & -\lambda & -\lambda \end{vmatrix}$$

$$= (1-\lambda) \begin{vmatrix} 1-\lambda & -1 \\ -\lambda & -\lambda \end{vmatrix} + \begin{vmatrix} -1 & 1 \\ -\lambda & -\lambda \end{vmatrix}$$

$$= -\lambda(1-\lambda) \begin{vmatrix} 1-\lambda & -1 \\ 1 & 1 \end{vmatrix} - \lambda \begin{vmatrix} -1 & 1 \\ 1 & 1 \end{vmatrix}$$

$$= -\lambda(1-\lambda)(2-\lambda) + 2\lambda$$

$$= -\lambda[(1-\lambda)(2-\lambda) - 2]$$

$$= -\lambda^2(\lambda - 3)$$

The roots of the characteristic equation, $-\lambda^2(\lambda - 3) = 0$, are $\lambda = 0$ and $\lambda = 3$; these are the eigenvalues of C. ∎

LINEAR ALGEBRA

Determining the Eigenvectors of a Matrix

In order to determine the eigenvectors of a matrix, you must first determine the eigenvalues. Substitute one eigenvalue λ into the equation $A\mathbf{x} = \lambda\mathbf{x}$—or, equivalently, into $(A - \lambda I)\mathbf{x} = \mathbf{0}$—and solve for $\mathbf{x}$; the resulting *nonzero* solutions form the set of eigenvectors of A corresponding to the selected eigenvalue. This process is then repeated for each of the remaining eigenvalues.

Example 3: Determine the eigenvectors of the matrix

$$A = \begin{bmatrix} 1 & -2 \\ 3 & -4 \end{bmatrix}$$

In Example 1, the eigenvalues of this matrix were found to be $\lambda = -1$ and $\lambda = -2$. Therefore, there are nonzero vectors $\mathbf{x}$ such that $A\mathbf{x} = -\mathbf{x}$ (the eigenvectors corresponding to the eigenvalue $\lambda = -1$), and there are nonzero vectors $\mathbf{x}$ such that $A\mathbf{x} = -2\mathbf{x}$ (the eigenvectors corresponding to the eigenvalue $\lambda = -2$). The eigenvectors corresponding to the eigenvalue $\lambda = -1$ are the solutions of the equation $A\mathbf{x} = -\mathbf{x}$:

$$\begin{bmatrix} 1 & -2 \\ 3 & -4 \end{bmatrix}\mathbf{x} = -\mathbf{x}$$

$$\begin{bmatrix} 1 & -2 \\ 3 & -4 \end{bmatrix}\begin{bmatrix} x_1 \\ x_2 \end{bmatrix} = -\begin{bmatrix} x_1 \\ x_2 \end{bmatrix}$$

This is equivalent to the pair of equations

$$x_1 - 2x_2 = -x_1$$
$$3x_1 - 4x_2 = -x_2$$

which simplifies to

$$2x_1 - 2x_2 = 0$$
$$3x_1 - 3x_2 = 0$$

[Note that these equations are not independent. If they *were* independent, then only $(x_1, x_2)^T = (0, 0)^T$ would satisfy them; this would signal that an error was made in the determination of the eigenvalues. If the eigenvalues are calculated correctly, then there *must* be nonzero solutions to each system $A\mathbf{x} = \lambda\mathbf{x}$.] The equations above are satisfied by all vectors $\mathbf{x} = (x_1, x_2)^T$ such that $x_2 = x_1$. Any such vector has the form $(x_1, x_1)^T$ and is therefore a multiple of the vector $(1, 1)^T$. Consequently, the eigenvectors of A corresponding to the eigenvalue $\lambda = -1$ are precisely the vectors

$$t\begin{bmatrix} 1 \\ 1 \end{bmatrix}$$

where t is any nonzero scalar.

The eigenvectors corresponding to the eigenvalue $\lambda = -2$ are the solutions of the equation $A\mathbf{x} = -2\mathbf{x}$:

$$\begin{bmatrix} 1 & -2 \\ 3 & -4 \end{bmatrix}\mathbf{x} = -2\mathbf{x}$$

$$\begin{bmatrix} 1 & -2 \\ 3 & -4 \end{bmatrix}\begin{bmatrix} x_1 \\ x_2 \end{bmatrix} = -2\begin{bmatrix} x_1 \\ x_2 \end{bmatrix}$$

This is equivalent to the "pair" of equations

$$3x_1 - 2x_2 = 0$$
$$3x_1 - 2x_2 = 0$$

Again, note that these equations are not independent. They are satisfied by any vector $\mathbf{x} = (x_1, x_2)^{\mathrm{T}}$ that is a multiple of the vector $(2, 3)^{\mathrm{T}}$; that is, the eigenvectors of A corresponding to the eigenvalue $\lambda = -2$ are the vectors

$$t \begin{bmatrix} 2 \\ 3 \end{bmatrix}$$

where t is any nonzero scalar. ■

Example 4: Consider the general 2×2 matrix

$$A = \begin{bmatrix} a & b \\ c & d \end{bmatrix}$$

(a) Express the eigenvalues of A in terms of a, b, c, and d. What can you say about the eigenvalues if $b = c$ (that is, if the matrix A is symmetric)?

(b) Verify that the sum of the eigenvalues is equal to the sum of the diagonal entries in A.

(c) Verify that the product of the eigenvalues is equal to the determinant of A.

(d) What can you say about the matrix A if one of its eigenvalues is 0?

The solutions are as follows:

(a) The eigenvalues of A are found by solving the characteristic equation, $\det(A - \lambda I) = 0$:

$$\det(A - \lambda I) = 0$$

$$\det \begin{bmatrix} a - \lambda & b \\ c & d - \lambda \end{bmatrix} = 0$$

$$(a-\lambda)(d-\lambda) - bc = 0$$
$$\lambda^2 - (a+d)\lambda + (ad - bc) = 0 \qquad (*)$$

The solutions of this equation—which are the eigenvalues of *A*—are found by using the quadratic formula:

$$\lambda = \frac{(a+d) \pm \sqrt{(a+d)^2 - 4(ad - bc)}}{2} \qquad (**)$$

The discriminant in (**) can be rewritten as follows:

$$(a+d)^2 - 4(ad - bc) = a^2 + 2ad + d^2 - 4ad + 4bc$$
$$= a^2 - 2ad + d^2 + 4bc$$
$$= (a-d)^2 + 4bc$$

Therefore, if $b = c$, the discriminant becomes $(a-d)^2 + 4b^2 = (a-d)^2 + (2b)^2$. Being the sum of two squares, this expression is nonnegative, so (**) implies that the eigenvalues are real. In fact, it can be shown that the eigenvalues of *any* real, symmetric matrix are real.

(b) The sum of the eigenvalues can be found by adding the two values expressed in (**) above:

$$\lambda_1 + \lambda_2 = \frac{(a+d) + \sqrt{(a+d)^2 - 4(ad - bc)}}{2}$$
$$+ \frac{(a+d) - \sqrt{(a+d)^2 - 4(ad - bc)}}{2}$$
$$= \frac{a+d}{2} + \frac{a+d}{2}$$
$$= a+d$$

which does indeed equal the sum of the diagonal entries of A. (The sum of the diagonal entries of any square matrix is called the **trace** of the matrix.) Another method for determining the sum of the eigenvalues, and one which works for any size matrix, is to examine the characteristic equation. From the theory of polynomial equations, it is known that if $p(\lambda)$ is a monic polynomial of degree n, then the sum of the roots of the equation $p(\lambda) = 0$ is the opposite of the coefficient of the λ^{n-1} term in $p(\lambda)$. The sum of the roots of equation (*) is therefore $-[-(a + d)] = a + d$, as desired. This second method can be used to prove that *the sum of the eigenvalues of any (square) matrix is equal to the trace of the matrix.*

(c) The product of the eigenvalues can be found by multiplying the two values expressed in (**) above:

$$\lambda_1 \lambda_2 = \left(\frac{(a+d) + \sqrt{(a+d)^2 - 4(ad - bc)}}{2} \right)$$

$$\times \left(\frac{(a+d) - \sqrt{(a+d)^2 - 4(ad - bc)}}{2} \right)$$

$$= \left(\frac{a+d}{2} \right)^2 - \left(\frac{\sqrt{(a+d)^2 - 4(ad - bc)}}{2} \right)^2$$

$$= \frac{(a+d)^2 - \left[(a+d)^2 - 4(ad - bc) \right]}{4}$$

$$= ad - bc$$

which is indeed equal to the determinant of A. Another proof that the product of the eigenvalues of *any* (square) matrix is equal to its determinant proceeds as follows. If A is an $n \times n$ matrix, then its characteristic polynomial, $p(\lambda)$, is monic of degree n. The equation $p(\lambda) = 0$ therefore has n roots: λ_1, λ_2, ..., λ_n (which may not be distinct); these are the eigenvalues. Consequently, the polynomial $p(\lambda) = \det(A - \lambda I)$ can be expressed in factored form as follows:

$$\det(A - \lambda I) = (\lambda_1 - \lambda)(\lambda_2 - \lambda) \cdots (\lambda_n - \lambda)$$

Substituting $\lambda = 0$ into this identity gives the desired result: $\det A = \lambda_1 \lambda_2 \cdots \lambda_n$.

(d) If 0 is an eigenvalue of a matrix A, then the equation $A\mathbf{x} = \lambda\mathbf{x} = 0\mathbf{x} = \mathbf{0}$ must have nonzero solutions, which are the eigenvectors associated with $\lambda = 0$. But if A is square and $A\mathbf{x} = \mathbf{0}$ has nonzero solutions, then A must be singular, that is, $\det A$ must be 0. This observation establishes the following fact: *Zero is an eigenvalue of a matrix if and only if the matrix is singular.* ■

Example 5: Determine the eigenvalues and eigenvectors of the identity matrix I without first calculating its characteristic equation.

The equation $A\mathbf{x} = \lambda\mathbf{x}$ characterizes the eigenvalues and associated eigenvectors of any matrix A. If $A = I$, this equation becomes $\mathbf{x} = \lambda\mathbf{x}$. Since $\mathbf{x} \neq \mathbf{0}$, this equation implies $\lambda = 1$; then, from $\mathbf{x} = 1\mathbf{x}$, every (nonzero) vector is an eigenvector of I. Remember the definition: $\mathbf{x}$ is an eigenvector of a matrix A if $A\mathbf{x}$ is a scalar multiple of $\mathbf{x}$ and $\mathbf{x} \neq \mathbf{0}$. Since multiplication by I leaves $\mathbf{x}$ unchanged, *every* (nonzero) vector must be an eigenvector of I, and the only possible scalar multiple— eigenvalue—is 1. ■

Example 6: The *Cayley-Hamilton Theorem* states that any square matrix satisfies its own characteristic equation; that is, if A has characteristic polynomial $p(\lambda)$, then $p(A) = 0$. To illustrate, consider the matrix

$$A = \begin{bmatrix} 1 & -2 \\ 3 & -4 \end{bmatrix}$$

from Example 1. Since its characteristic polynomial is $p(\lambda) = \lambda^2 + 3\lambda + 2$, the Cayley-Hamilton Theorem states that $p(A)$ should equal the zero matrix, 0. This is verified as follows:

$$p(A) = A^2 + 3A + 2I$$

$$= \begin{bmatrix} 1 & -2 \\ 3 & -4 \end{bmatrix}^2 + 3 \begin{bmatrix} 1 & -2 \\ 3 & -4 \end{bmatrix} + 2 \begin{bmatrix} 1 & 0 \\ 0 & 1 \end{bmatrix}$$

$$= \begin{bmatrix} -5 & 6 \\ -9 & 10 \end{bmatrix} + \begin{bmatrix} 3 & -6 \\ 9 & -12 \end{bmatrix} + \begin{bmatrix} 2 & 0 \\ 0 & 2 \end{bmatrix}$$

$$= \begin{bmatrix} 0 & 0 \\ 0 & 0 \end{bmatrix}$$

$$= 0 \quad \checkmark$$

If A is an n by n matrix, then its characteristic polynomial has degree n. The Cayley-Hamilton Theorem then provides a way to express every integer power A^k in terms of a polynomial in A of degree less than n. For example, for the 2×2 matrix above, the fact that $A^2 + 3A + 2I = 0$ implies $A^2 = -3A - 2I$. Thus, A^2 is expressed in terms of a polynomial of degree 1 in A. Now, by repeated applications, *every* positive integer power of this 2 by 2 matrix A can be expressed as a polynomial of degree less than 2. To illustrate, note the following calculation for expressing A^5 in terms of a linear polynomial in A; the key is to consistently replace A^2 by $-3A - 2I$ and simplify:

$$
\begin{aligned}
A^5 &= A^2 \cdot A^2 \cdot A \\
&= (-3A - 2I) \cdot (-3A - 2I) \cdot A \\
&= (9A^2 + 12A + 4I) \cdot A \\
&= [9(-3A - 2I) + 12A + 4I] \cdot A \\
&= (-15A - 14I) \cdot A \\
&= -15A^2 - 14A \\
&= -15(-3A - 2I) - 14A \\
&= 31A + 30I
\end{aligned}
$$

This result yields

$$
A^5 = 31A + 30I = 31\begin{bmatrix} 1 & -2 \\ 3 & -4 \end{bmatrix} + 30\begin{bmatrix} 1 & 0 \\ 0 & 1 \end{bmatrix} = \begin{bmatrix} 61 & -62 \\ 93 & -94 \end{bmatrix}
$$

a calculation which you are welcome to verify be performing the repeated multiplications

$$
A^5 = \begin{bmatrix} 1 & -2 \\ 3 & -4 \end{bmatrix}\begin{bmatrix} 1 & -2 \\ 3 & -4 \end{bmatrix}\begin{bmatrix} 1 & -2 \\ 3 & -4 \end{bmatrix}\begin{bmatrix} 1 & -2 \\ 3 & -4 \end{bmatrix}\begin{bmatrix} 1 & -2 \\ 3 & -4 \end{bmatrix}
$$

The Cayley-Hamilton Theorem can also be used to express the inverse of an invertible matrix A as a polynomial in A. For example, for the 2 by 2 matrix A above,

$$
\begin{aligned}
A^2 + 3A + 2I &= 0 \\
A^2 + 3A &= -2I \\
A(A + 3I) &= -2I \\
A \cdot \left[-\tfrac{1}{2}(A + 3I) \right] &= I \\
A^{-1} &= -\tfrac{1}{2}(A + 3I) \qquad (*)
\end{aligned}
$$

This result can be easily verified. The inverse of an invertible 2 by 2 matrix is found by first interchanging the entries on the diagonal, then taking the opposite of the each off-diagonal entry, and, finally, dividing by the determinant of A. Since det $A = 2$,

$$A = \begin{bmatrix} 1 & -2 \\ 3 & -4 \end{bmatrix} \Rightarrow A^{-1} = \tfrac{1}{2}\begin{bmatrix} -4 & 2 \\ -3 & 1 \end{bmatrix} = \begin{bmatrix} -2 & 1 \\ -\tfrac{3}{2} & \tfrac{1}{2} \end{bmatrix}$$

but

$$-\tfrac{1}{2}(A+3I) = -\tfrac{1}{2}\left(\begin{bmatrix} 1 & -2 \\ 3 & -4 \end{bmatrix} + 3\begin{bmatrix} 1 & 0 \\ 0 & 1 \end{bmatrix}\right) = -\tfrac{1}{2}\begin{bmatrix} 4 & -2 \\ 3 & -1 \end{bmatrix} = \begin{bmatrix} -2 & 1 \\ -\tfrac{3}{2} & \tfrac{1}{2} \end{bmatrix}$$

validating the expression in (*) for A^{-1}. The same ideas used to express any positive integer power of an n by n matrix A in terms of a polynomial of degree less than n can also be used to express any *negative* integer power of (an invertible matrix) A in terms of such a polynomial. ∎

Example 7: Let A be a square matrix. How do the eigenvalues and associated eigenvectors of A^2 compare with those of A? Assuming that A is invertible, how do the eigenvalues and associated eigenvectors of A^{-1} compare with those of A?

Let λ be an eigenvalue of the matrix A, and let $\mathbf{x}$ be a corresponding eigenvector. Then $A\mathbf{x} = \lambda\mathbf{x}$, and it follows from this equation that

$$A^2\mathbf{x} = A(A\mathbf{x}) = A(\lambda\mathbf{x}) = \lambda(A\mathbf{x}) = \lambda(\lambda\mathbf{x}) = \lambda^2\mathbf{x}$$

Therefore, λ^2 is an eigenvalue of A^2, and $\mathbf{x}$ is the corresponding eigenvector. Now, if A is invertible, then A has no zero eigenvalues, and the following calculations are justified:

$$Ax = \lambda x$$
$$A^{-1}(Ax) = A^{-1}(\lambda x)$$
$$x = \lambda(A^{-1}x)$$
$$\lambda^{-1}x = A^{-1}x$$

so λ^{-1} is an eigenvalue of A^{-1} with corresponding eigenvector x. ■

Eigenspaces

Let A be an $n \times n$ matrix and consider the set $E = \{x \in \mathbf{R}^n : Ax = \lambda x\}$. If $x \in E$, then so is tx for any scalar t, since

$$A(tx) = t(Ax) = t(\lambda x) = \lambda(tx) \quad \Rightarrow \quad tx \in E$$

Furthermore, if x_1 and x_2 are in E, then

$$A(x_1 + x_2) = Ax_1 + Ax_2 = \lambda x_1 + \lambda x_2$$
$$= \lambda(x_1 + x_2) \quad \Rightarrow \quad x_1 + x_2 \in E$$

These calculations show that E is closed under scalar multiplication and vector addition, so E is a subspace of $\mathbf{R}^n$. Clearly, the zero vector belongs to E; but more notably, *the nonzero elements in E are precisely the eigenvectors of A corresponding to the eigenvalue λ*. When the zero vector is adjoined to the collection of eigenvectors corresponding to a particular eigenvalue, the resulting collection,

$$\left\{ \begin{array}{c} \text{eigenvectors of } A \text{ corresponding} \\ \text{to the eigenvalue } \lambda \end{array} \right\} \cup \{\mathbf{0}\}$$

forms a vector space called the **eigenspace** of A corresponding to the eigenvalue λ. Since it depends on both A and the selection of one of its eigenvalues, the notation

$$E_\lambda(A) = \{\mathbf{x}: A\mathbf{x} = \lambda\mathbf{x}\}$$

will be used to denote this space. Since the equation $A\mathbf{x} = \lambda\mathbf{x}$ is equivalent to $(A - \lambda I)\mathbf{x} = \mathbf{0}$, the eigenspace $E_\lambda(A)$ can also be characterized as the nullspace of $A - \lambda I$:

$$E_\lambda(A) = \{\mathbf{x}: A\mathbf{x} = \lambda\mathbf{x}\} = \{\mathbf{x}: (A-\lambda I)\mathbf{x} = \mathbf{0}\} = N(A-\lambda I)$$

This observation provides an immediate proof that $E_\lambda(A)$ is a subspace of $\mathbf{R}^n$.

Recall the matrix

$$A = \begin{bmatrix} 1 & -2 \\ 3 & -4 \end{bmatrix}$$

given in Example 2 above. The determination of the eigenvectors of A shows that its eigenspaces are

$$E_{-1}(A) = \left\{\mathbf{x} \in \mathbf{R}^2: \mathbf{x} = t\begin{bmatrix} 1 \\ 1 \end{bmatrix}, t \in \mathbf{R}\right\}$$

and

$$E_{-2}(A) = \left\{\mathbf{x} \in \mathbf{R}^2: \mathbf{x} = t\begin{bmatrix} 2 \\ 3 \end{bmatrix}, t \in \mathbf{R}\right\}$$

$E_{-1}(A)$ is the line in $\mathbf{R}^2$ through the origin and the point $(1, 1)$, and $E_{-2}(A)$ is the line through the origin and the point $(2, 3)$. Both of these eigenspaces are 1-dimensional subspaces of $\mathbf{R}^2$.

Example 8: Determine the eigenspaces of the matrix

$$B = \begin{bmatrix} 1 & 0 & 2 \\ 0 & 3 & 0 \\ 2 & 0 & 1 \end{bmatrix}$$

First, form the matrix

$$B - \lambda I = \begin{bmatrix} 1-\lambda & 0 & 2 \\ 0 & 3-\lambda & 0 \\ 2 & 0 & 1-\lambda \end{bmatrix} \qquad (*)$$

The determinant will be computed by performing a Laplace expansion along the second row:

$$\det(B - \lambda I) = \det \begin{bmatrix} 1-\lambda & 0 & 2 \\ 0 & 3-\lambda & 0 \\ 2 & 0 & 1-\lambda \end{bmatrix}$$

$$= (3-\lambda) \begin{vmatrix} 1-\lambda & 2 \\ 2 & 1-\lambda \end{vmatrix}$$

$$= (3-\lambda)\left[(1-\lambda)^2 - 2^2\right]$$

$$= (3-\lambda)\left[(1-\lambda) + 2\right]\left[(1-\lambda) - 2\right]$$

$$= (3-\lambda)(3-\lambda)(-1-\lambda)$$

The roots of the characteristic equation,

$$(3-\lambda)(3-\lambda)(-1-\lambda) = 0$$

are clearly $\lambda = -1$ and 3, with 3 being a double root; these are the eigenvalues of B. The associated eigenvectors can now be found. Substituting $\lambda = -1$ into the matrix $B - \lambda I$ in (*) gives

$$(B - \lambda I)_{\lambda = -1} = \begin{bmatrix} 1-\lambda & 0 & 2 \\ 0 & 3-\lambda & 0 \\ 2 & 0 & 1-\lambda \end{bmatrix}_{\lambda = -1} = \begin{bmatrix} 2 & 0 & 2 \\ 0 & 4 & 0 \\ 2 & 0 & 2 \end{bmatrix}$$

which is the coefficient matrix for the equation $(B - \lambda I)\mathbf{x} = \mathbf{0}$ with $\lambda = -1$, which determines the eigenvectors corresponding to the eigenvalue $\lambda = -1$. These eigenvectors are the nonzero solutions of

$$\begin{bmatrix} 2 & 0 & 2 \\ 0 & 4 & 0 \\ 2 & 0 & 2 \end{bmatrix}\begin{bmatrix} x_1 \\ x_2 \\ x_3 \end{bmatrix} = \begin{bmatrix} 0 \\ 0 \\ 0 \end{bmatrix} \Rightarrow \begin{array}{r} 2x_1 + 2x_3 = 0 \\ 4x_2 = 0 \\ 2x_1 + 2x_3 = 0 \end{array}$$

The identical first and third equations imply that $x_1 + x_3 = 0$— that is, $x_3 = -x_1$—and the second equation says $x_2 = 0$. Therefore, the eigenvectors of B associated with the eigenvalue $\lambda = -1$ are all vectors of the form $(x_1, 0, -x_1)^T = x_1(1, 0, -1)^T$ for $x_1 \neq 0$. Removing the restriction that the scalar multiple be nonzero includes the zero vector and gives the full eigenspace:

$$E_{-1}(B) = \left\{ \mathbf{x} \in \mathbf{R}^3 : \mathbf{x} = t\begin{bmatrix} 1 \\ 0 \\ -1 \end{bmatrix}, \; t \in \mathbf{R} \right\}$$

Now, since

$$(B - \lambda I)_{\lambda = 3} = \begin{bmatrix} 1-\lambda & 0 & 2 \\ 0 & 3-\lambda & 0 \\ 2 & 0 & 1-\lambda \end{bmatrix}_{\lambda = 3} = \begin{bmatrix} -2 & 0 & 2 \\ 0 & 0 & 0 \\ 2 & 0 & -2 \end{bmatrix}$$

the eigenvectors corresponding to the eigenvalue $\lambda = 3$ are the nonzero solutions of

$$\begin{bmatrix} -2 & 0 & 2 \\ 0 & 0 & 0 \\ 2 & 0 & -2 \end{bmatrix}\begin{bmatrix} x_1 \\ x_2 \\ x_3 \end{bmatrix} = \begin{bmatrix} 0 \\ 0 \\ 0 \end{bmatrix} \implies \begin{array}{c} -2x_1 + 2x_3 = 0 \\ 2x_1 - 2x_3 = 0 \end{array}$$

These equations imply that $x_3 = x_1$, and since there is no restriction on x_2, this component is arbitrary. Therefore, the eigenvectors of B associated with $\lambda = 3$ are all nonzero vectors of the form $(x_1, x_2, x_1)^T = x_1(1, 0, 1)^T + x_2(0, 1, 0)^T$. The inclusion of the zero vector gives the eigenspace:

$$E_3(B) = \left\{ \mathbf{x} \in \mathbf{R}^3 : \mathbf{x} = t_1 \begin{bmatrix} 1 \\ 0 \\ 1 \end{bmatrix} + t_2 \begin{bmatrix} 0 \\ 1 \\ 0 \end{bmatrix}; \ t_1, t_2 \in \mathbf{R} \right\}$$

Note that $\dim E_{-1}(B) = 1$ and $\dim E_3(B) = 2$. ∎

Diagonalization

First, a theorem:

Theorem O. Let A be an n by n matrix. If the n eigenvalues of A are distinct, then the corresponding eigenvectors are linearly independent.

Proof. The proof of this theorem will be presented explicitly for $n = 2$; the proof in the general case can be constructed based on the same method. Therefore, let A be 2 by 2, and denote its eigenvalues by λ_1 and λ_2 and the corresponding eigenvectors by $\mathbf{v}_1$ and $\mathbf{v}_2$ (so that $A\mathbf{v}_1 = \lambda_1\mathbf{v}_1$ and $A\mathbf{v}_2 = \lambda_2\mathbf{v}_2$). The goal is to prove that if $\lambda_1 \neq \lambda_2$, then $\mathbf{v}_1$ and $\mathbf{v}_2$ are linearly independent. Assume that

$$c_1 \mathbf{v}_1 + c_2 \mathbf{v}_2 = \mathbf{0} \quad (*)$$

is a linear combination of $\mathbf{v}_1$ and $\mathbf{v}_2$ that gives the zero vector; the goal is to show that the above equation implies that c_1 and c_2 must be zero. First, multiply both sides of (*) by the matrix A:

$$A(c_1 \mathbf{v}_1 + c_2 \mathbf{v}_2) = c_1(A\mathbf{v}_1) + c_2(A\mathbf{v}_2) = \mathbf{0}$$

Next, use the fact that $A\mathbf{v}_1 = \lambda \mathbf{v}_1$ and $A\mathbf{v}_2 = \lambda \mathbf{v}_2$ to write

$$c_1(\lambda_1 \mathbf{v}_1) + c_2(\lambda_2 \mathbf{v}_2) = \mathbf{0} \quad (**)$$

Now, multiply both sides of (*) by λ_2 and subtract the resulting equation, $c_1 \lambda_2 \mathbf{v}_1 + c_2 \lambda_2 \mathbf{v}_2 = \mathbf{0}$, from (**):

$$c_1(\lambda_1 - \lambda_2)\mathbf{v}_1 = \mathbf{0}$$

Since the eigenvalues are distinct, $\lambda_1 - \lambda_2 \neq 0$, and since $\mathbf{v}_1 \neq \mathbf{0}$ ($\mathbf{v}_1$ is an eigenvector), this last equation implies that $c_1 = 0$. Multiplying both sides of (*) by λ_1 and subtracting the resulting equation from (**) leads to $c_2(\lambda_2 - \lambda_1)\mathbf{v}_2 = \mathbf{0}$ and then, by the same reasoning, to the conclusion that $c_2 = 0$ also. ∎

Using the same notation as in the proof of Theorem O, assume that A is a 2 by 2 matrix with distinct eigenvalues and form the matrix

$$V = \begin{bmatrix} | & | \\ \mathbf{v}_1 & \mathbf{v}_2 \\ | & | \end{bmatrix}$$

whose columns are the eigenvectors of A. Now consider the product AV; since $A\mathbf{v}_1 = \lambda_1 \mathbf{v}_1$ and $A\mathbf{v}_2 = \lambda_2 \mathbf{v}_2$,

$$AV = A \begin{bmatrix} | & | \\ \mathbf{v}_1 & \mathbf{v}_2 \\ | & | \end{bmatrix} = \begin{bmatrix} | & | \\ A\mathbf{v}_1 & A\mathbf{v}_2 \\ | & | \end{bmatrix} = \begin{bmatrix} | & | \\ \lambda_1\mathbf{v}_1 & \lambda_2\mathbf{v}_2 \\ | & | \end{bmatrix} \quad (*)$$

This last matrix can be expressed as the following product:

$$\begin{bmatrix} | & | \\ \lambda_1\mathbf{v}_1 & \lambda_2\mathbf{v}_2 \\ | & | \end{bmatrix} = \begin{bmatrix} | & | \\ \mathbf{v}_1 & \mathbf{v}_2 \\ | & | \end{bmatrix} \begin{bmatrix} \lambda_1 & \\ & \lambda_2 \end{bmatrix} \quad (**)$$

If Λ denotes the diagonal matrix whose entries are the eigenvalues of A,

$$\Lambda = \begin{bmatrix} \lambda_1 & \\ & \lambda_2 \end{bmatrix}$$

then equations (*) and (**) together imply $AV = V\Lambda$. If $\mathbf{v}_1$ and $\mathbf{v}_2$ are linearly independent, then the matrix V is invertible. Form the matrix V^{-1} and left multiply both sides of the equation $AV = V\Lambda$ by V^{-1}:

$$V^{-1}AV = \Lambda = \begin{bmatrix} \lambda_1 & \\ & \lambda_2 \end{bmatrix}$$

(Although this calculation has been shown for $n = 2$, it clearly can be applied to an n by n matrix of any size.) This process of forming the product $V^{-1}AV$, resulting in the diagonal matrix Λ of its eigenvalues, is known as the **diagonalization** of the matrix A, and the matrix of eigenvectors, V, is said to **diagonalize** A. *The key to diagonalizing an n by n matrix A is the ability to form the n by n eigenvector matrix V and its inverse; this requires a full set of n linearly independent eigenvectors.* A sufficient (but not necessary) condition that will guarantee that this requirement is fulfilled is provided by Theorem O: if the *n* by *n* matrix *A* has *n distinct* eigenvalues.

One useful application of diagonalization is to provide a simple way to express integer powers of the matrix *A*. If *A* can be diagonalized, then $V^{-1}AV = \Lambda$, which implies

$$A = V\Lambda V^{-1}$$

When expressed in this form, it is easy to form integer powers of *A*. For example, if *k* is a positive integer, then

$$A^k = (V\Lambda V^{-1})^k = \underbrace{(V\Lambda V^{-1})\cdot(V\Lambda V^{-1})\cdots(V\Lambda V^{-1})\cdot(V\Lambda V^{-1})}_{k \text{ factors}}$$

$$= V\Lambda\underbrace{(V^{-1}V)\cdot\Lambda(V^{-1}V)\cdots\Lambda(V^{-1}V)\cdot\Lambda}_{k \text{ factors}}V^{-1}$$

$$= V\Lambda^k V^{-1}$$

The power Λ^k is trivial to compute: If $\lambda_1, \lambda_2, \ldots, \lambda_n$ are the entries of the diagonal matrix Λ, then Λ^k is diagonal with entries $\lambda_1^k, \lambda_2^k, \ldots, \lambda_n^k$. Therefore,

$$A^k = V \begin{bmatrix} \lambda_1^k & & & \\ & \lambda_2^k & & \\ & & \ddots & \\ & & & \lambda_n^k \end{bmatrix} V^{-1}$$

Example 9: Compute A^{10} for the matrix

$$A = \begin{bmatrix} 1 & -2 \\ 3 & -4 \end{bmatrix}$$

This is the matrix of Example 1. Its eigenvalues are $\lambda_1 = -1$ and $\lambda_2 = -2$, with corresponding eigenvectors $\mathbf{v}_1 = (1, 1)^T$ and $\mathbf{v}_2 = (2, 3)^T$. Since these eigenvectors are linearly independent (which was to be expected, since the eigenvalues are distinct), the eigenvector matrix V has an inverse,

$$V = \begin{bmatrix} 1 & 2 \\ 1 & 3 \end{bmatrix} \;\Rightarrow\; V^{-1} = \begin{bmatrix} 3 & -2 \\ -1 & 1 \end{bmatrix}$$

Thus, A can be diagonalized, and the diagonal matrix $\Lambda = V^{-1}AV$ is

$$\Lambda = \begin{bmatrix} \lambda_1 & \\ & \lambda_2 \end{bmatrix} = \begin{bmatrix} -1 & \\ & -2 \end{bmatrix}$$

Therefore,

$$
\begin{aligned}
A^{10} &= (V\Lambda V^{-1})^{10} \\
&= V\Lambda^{10}V^{-1} \\
&= \begin{bmatrix} 1 & 2 \\ 1 & 3 \end{bmatrix}\begin{bmatrix} (-1)^{10} & \\ & (-2)^{10} \end{bmatrix}\begin{bmatrix} 3 & -2 \\ -1 & 1 \end{bmatrix} \\
&= \begin{bmatrix} 1 & 2 \\ 1 & 3 \end{bmatrix}\begin{bmatrix} 1 & \\ & 1024 \end{bmatrix}\begin{bmatrix} 3 & -2 \\ -1 & 1 \end{bmatrix} \\
&= \begin{bmatrix} 1 & 2\cdot 1024 \\ 1 & 3\cdot 1024 \end{bmatrix}\begin{bmatrix} 3 & -2 \\ -1 & 1 \end{bmatrix} \\
&= \begin{bmatrix} 3-2\cdot 1024 & -2+2\cdot 1024 \\ 3-3\cdot 1024 & -2+3\cdot 1024 \end{bmatrix} \\
&= \begin{bmatrix} -2045 & 2046 \\ -3069 & 3070 \end{bmatrix}
\end{aligned}
$$

∎

Although an n by n matrix with n distinct eigenvalues is guaranteed to be diagonalizable, an n by n matrix that does not have n distinct eigenvalues may still be diagonalizable. If the eigenspace corresponding to each k-fold root λ of the characteristic equation is k dimensional, then the matrix will be diagonalizable. In other words, diagonalization is guaranteed if the *geometric* multiplicity of each eigenvalue (that is, the dimension of its corresponding eigenspace) matches its *algebraic* multiplicity (that is, its multiplicity as a root of the characteristic equation). Here's an illustration of this result. The 3 by 3 matrix

$$B = \begin{bmatrix} 1 & 0 & 2 \\ 0 & 3 & 0 \\ 2 & 0 & 1 \end{bmatrix}$$

of Example 8 has just two eigenvalues: $\lambda_1 = -1$ and $\lambda_2 = 3$. The algebraic multiplicity of the eigenvalue $\lambda_1 = -1$ is one, and its corresponding eigenspace, $E_{-1}(B)$, is one dimensional. Furthermore, the algebraic multiplicity of the eigenvalue $\lambda_2 = 3$ is two, and its corresponding eigenspace, $E_3(B)$, is two dimensional. Therefore, the geometric multiplicities of the eigenvalues of B match their algebraic multiplicities. The conclusion, then, is that although the 3 by 3 matrix B does not have 3 distinct eigenvalues, it is nevertheless diagonalizable.

Here's the verification: Since $\{(1, 0, -1)^T\}$ is a basis for the 1-dimensional eigenspace corresponding to the eigenvalue $\lambda_1 = -1$, and $\{(0, 1, 0)^T, (1, 0, 1)^T\}$ is a basis for the 2-dimensional eigenspace corresponding to the eigenvalue $\lambda_2 = 3$, the matrix of eigenvectors reads

$$V = \begin{bmatrix} 1 & 0 & 1 \\ 0 & 1 & 0 \\ -1 & 0 & 1 \end{bmatrix}$$

Since the key to the diagonalization of the original matrix B is the invertibility of this matrix, V, evaluate det V and check that it is nonzero. Because det $V = 2$, the matrix V *is* invertible,

$$V^{-1} = \begin{bmatrix} \frac{1}{2} & 0 & -\frac{1}{2} \\ 0 & 1 & 0 \\ \frac{1}{2} & 0 & \frac{1}{2} \end{bmatrix}$$

so B is indeed diagonalizable:

$$V^{-1}BV = \begin{bmatrix} \frac{1}{2} & 0 & -\frac{1}{2} \\ 0 & 1 & 0 \\ \frac{1}{2} & 0 & \frac{1}{2} \end{bmatrix}\begin{bmatrix} 1 & 0 & 2 \\ 0 & 3 & 0 \\ 2 & 0 & 1 \end{bmatrix}\begin{bmatrix} 1 & 0 & 1 \\ 0 & 1 & 0 \\ -1 & 0 & 1 \end{bmatrix} = \begin{bmatrix} -1 & & \\ & 3 & \\ & & 3 \end{bmatrix} = \Lambda$$

Example 10: Diagonalize the matrix

$$A = \begin{bmatrix} 2 & -1 \\ -3 & 4 \end{bmatrix}$$

First, find the eigenvalues; since

$$\det(A - \lambda I) = \begin{vmatrix} 2-\lambda & -1 \\ -3 & 4-\lambda \end{vmatrix} = (2-\lambda)(4-\lambda) - 3$$
$$= \lambda^2 - 6\lambda + 5$$
$$= (\lambda - 1)(\lambda - 5)$$

the eigenvalues are $\lambda = 1$ and $\lambda = 5$. Because the eigenvalues are distinct, A is diagonalizable. Verify that an eigenvector corresponding to $\lambda = 1$ is $\mathbf{v}_1 = (1, 1)^T$, and an eigenvector corresponding to $\lambda = 5$ is $\mathbf{v}_2 = (1, -3)^T$. Therefore, the diagonalizing matrix is

$$V = \begin{bmatrix} | & | \\ \mathbf{v}_1 & \mathbf{v}_2 \\ | & | \end{bmatrix} = \begin{bmatrix} 1 & 1 \\ 1 & -3 \end{bmatrix}$$

and

$$\Lambda = \begin{bmatrix} \lambda_1 & 0 \\ 0 & \lambda_2 \end{bmatrix} = \begin{bmatrix} 1 & 0 \\ 0 & 5 \end{bmatrix}$$

Another application of diagonalization is in the construction of simple representative matrices for linear operators. Let A be the matrix defined above and consider the linear operator on $\mathbf{R}^2$ given by $T(\mathbf{x}) = A\mathbf{x}$. In terms of the nonstandard basis $B = \{\mathbf{v}_1 = (1, 1)^T, \mathbf{v}_2 = (1, -3)^T\}$ for $\mathbf{R}^2$, the matrix of T relative to B is Λ. Review Example 16 on pages 270–271.
■